D0982023

Effective Risk Communication

FOOD MICROBIOLOGY AND FOOD SAFETY SERIES

Food Microbiology and Food Safety publishes valuable, practical, and timely resources for professionals and researchers working on microbiological topics associated with foods, as well as food safety issues and problems.

Titles

Effective Risk Communication: A Message-Centered Approach, Timothy L. Sellnow, Robert R. Ulmer, Matthew W. Seeger, Robert S. Littlefield (Eds.) (2009)

Food Safety Culture, Frank Yiannas (2009)

Molecular Techniques in the Microbial Ecology of Fermented Foods, Luca Cocolin and Danilo Ercolini (Eds.) (2008)

Viruses in Foods, Sagar M. Goyal (Ed.) (2006)

Foodborne Parasites, Ynes R. Ortega (Ed.) (2006)

PCR Methods in Foods, John Maurer (Ed.) (2006)

Timothy L. Sellnow • Robert R. Ulmer •
Matthew W. Seeger • Robert S. Littlefield

Effective Risk Communication

A Message-Centered Approach

Timothy L. Sellnow, Ph.D.
University of Kentucky
Dept. Communication
249 Grehan Building
Lexington
KY 40506-0042
USA

Robert R. Ulmer, Ph.D.
University of Arkansas
Little Rock
Dept. Speech Communication
2801 South University
Little Rock AR 72204–1099
USA

Matthew W. Seeger, Ph.D.
Wayne State University
906 West Warren
Detroit MI 48202
USA

Robert S. Littlefield, Ph.D.
North Dakota State University
Dept. Communication
P. O. Box 5075
Fargo ND 58105
USA

ISBN 978-0-387-79726-7 e-ISBN 978-0-387-79727-4

DOI: 10.1007/978-0-387-79727-4

Library of Congress Control Number: 2008936988

Printed on acid-free paper

9 8 7 6 5 4 3 2 1

springer.com

Acknowledgments

The foundation for this work coincides with the 2003 launching of the National Center for Food Protection and Defense (NCFPD), a Department of Homeland Security Center of Excellence. Throughout the planning, implementation, and evolution of the NCFPD, risk communication has been a central feature. We are grateful to Will Hueston, Mike Osterholm, Frank Busta, and Shaun Kennedy for recognizing the importance of risk communication and for guiding and supporting our work. The NCFPD has consistently created opportunities for leading scholars in risk communication to gather, debate best practices, conduct evaluations, and offer recommendations. As such, we appreciate the participation and influence of Nick Alexander, Kris Boone, Lisa Brienzo, Tony Flood, Bob Gravani, Maria Lapinski, Julie Novak, Peter Sandman, Shari Veil, and Steven Venette. We have also had the good fortune to work with a number of promising graduate students on this project. In particular we thank Devon Wood, Will Whiting, Kathleen Vidoloff, Jennifer Reierson, and Elizabeth Petrun for their work with NCFPD case studies, including those found in this book. We also extend our thanks to Kimberly Cowden, Kelly Wolf, Nicole Dobransky, Elizabeth Webb, and Alyssa Millner for their contributions. As always, we extend our thanks to Chandice Johnson whose skills as an editor are second to none.

This research was supported by the U.S. Department of Homeland Security (Grant number N-00014-04-1-0659), through a grant awarded to the National Center for Food Protection and Defense at the University of Minnesota. Any opinions, findings, conclusions, or recommendations expressed in this publication are those of the author(s) and do not represent the policy or position of the Department of Homeland Security.

Contents

Part I
Conceptualizing a Message-Centered Approach to Risk Communication

Chapter 1
Introducing a Message-Centered Approach to Risk Communication

> *The latency phase of risk threats is coming to an end. The invisible hazards are becoming visible. Damage to and destruction of nature no longer occur outside our personal experience in the sphere of chemical, physical or biological chains of effects; instead they strike more and more clearly our eyes, ears and noses.—Ulrich Beck, Risk Society: Towards a New Modernity* (Beck, 1992, p. 55)

In its most basic form, risk is the absence of certainty. If we are absolutely certain of the results an action will produce, that action has no risk. In reality, we rarely, if ever, have the luxury of absolute certainty. Uncertainty, therefore, is the "central variable" in the risk perception and communication process (Palenchar & Heath, 2002, p. 131). In the absence of certainty, we must calculate the likely outcome of our activities based on the available information. From this perspective, risk is neither good nor bad. Rather, risk is a fundamental part of life. The way we manage risk, however, has a profound impact on our quality of life. As Ulrich Beck (1992) notes in the opening quotation, we are evolving into a society with increasingly acute levels of risk.

Advancing technology, unprecedented globalization, and the insatiable demand for energy are factors, among many others, that continue to complicate human activity and in so doing increase our uncertainty and risk. Mitroff and Anagnos (2001) observe that, over the past century, 28 major industry accidents causing 50 deaths or more have occurred world-wide. Mitroff and Anagnos point out that "the most disturbing part of this statistic is that half" of these catastrophic accidents "have occurred in the past fifteen or so years" (p. 3). Similar increases can be seen in nearly all aspects of life. Rather than making life more predictable, growth in technology, services, and population have increased our risks.

Although levels of risk continue to expand in complexity and intensity, our objective in this book is certainly not to slow down science. Rather, as Scherer and Juanillo (2003) aptly note, "What is really at issue is whether we have sufficient space to understand and talk about the specific prisms through which both scientists/experts and the public view risk" (p. 222). By characterizing the systematic development of risk messages, this book offers an assessment of current risk communication strategies and recommendations for improving the access and accuracy of risk messages for the general public in all settings.

Chapter 1 begins with a definition of risk communication. We then establish the importance of interaction and consider the impact of multiple messages in risk communication. Finally, we introduce the perspective of interacting arguments as a means for understanding, evaluating, and improving risk communication.

T. L. Sellnow et al. *Effective Risk Communication*

DOI: 10.1007/978-0-387-79727-4_1

Risk Communication	Crisis Communication
Risk centered: Projection about some harm occurring at some future date	Event centered: Specific incident that has occurred and produced harm
Messages regarding known probabilities of negative consequences and how they may be reduced	Messages regarding current state or conditions: Magnitude, immediacy, duration, control/remediation, cause, blame, consequences
Based on what is currently known	Based on what is known and what is not known
Long-term (pre-crisis stage) Message preparation (i.e., campaigns)	Short term (crisis stage) Less preparation (i.e., responsive)
Technical experts, scientists	Authority figure, emergency managers, technical experts
Personal scope	Community or regional scope
Mediated: Commercials, ads, brochures, pamphlets	Mediated: Press conferences, press releases, speeches, Web sites
Controlled and structured	Spontaneous and reactive

Fig. 1.1 Distinguishing features of risk communication and crisis communication. From Seeger et al. (2003)

Distinguishing Between Risk and Crisis

The ultimate purpose of risk communication is to avoid crises. By recognizing the uncertainty of risk situations, we are better able to determine the wisest and safest course of action. The ultimate result of our inability or failure to recognize and act upon risk is crisis. Crises are catastrophic events resulting in physical, emotional, or financial harm. Although the terms risk communication and crisis communication are often used interchangeably, there are clear distinctions between them. As Fig. 1.1 demonstrates, risk messages emerge long before a crisis event occurs. In fact, the ultimate goal of honest and effective risk communication is to avoid a crisis event. Hence, risk messages are typically forward-looking in hope of reducing the likelihood of a crisis event in the long-term. The evidence used in risk messages is based on information from technical experts and is adapted for audience members to consider on a personal level. Like AIDS awareness campaigns, for example, many risk communication campaigns are carefully controlled and orchestrated. Conversely, crisis communication takes place during and in the wake of the actual event. Crisis responders focus their communication on the events at hand and on what must be done immediately to resolve or contain the crisis. Once the crisis has passed, communication shifts back to understanding what went wrong and how the risk of future crises can be limited. In short, crisis communication focuses on containing and recovering from a dangerous event. Conversely, risk communication seeks to influence behavior and policies so that a crisis situation can be averted.

A Working Definition of Risk Communication

Risk communication as an area of investigation "grew out of risk perception and risk management studies" (Heath & Palenchar, 2000, p. 134). The ultimate goal of risk communication research is to "increase the quality of risk decisions through better communications" (Palenchar & Heath, 2002, p. 129). The need for such improvement is glaring. For example, government agencies have a long history of a "public information model of communication that stresses the one-way dissemination of information" (McComas, 2003, p. 166). This linear view of risk communication fails to solicit feedback from those who are asked to tolerate prescribed risk levels. Thus, in the linear view of risk communication, the potential for abuse or discrimination increases.

A notable turning point in this tendency to emphasize a one-way form of risk communication occurred in 1983 when the National Research Council (NRC) completed an extensive study of risk assessment by government agencies. The report, "Risk Assessment in the Federal Government: Managing the Process," stressed that risk communication is a key component in the risk assessment process. Moreover, the report revealed a void in risk communication research. In response to this void, the NRC formed the Committee on Risk Perception and Communication. This committee published the influential book, *Improving Risk Communication*, in 1989. With this publication, NRC established risk communication as a "democratic dialogue" (1989, p. 21). Specifically, it proposed the following definition:

> Risk communication is an interactive process of exchange of information and opinion among individuals, groups, and institutions. It involves multiple messages about the nature of risk and other messages, not strictly about risk, that express concerns, opinions, or reaction to risk messages or to legal or institutional arrangements for risk management. (p. 21)

This definition makes two influential contributions to our understanding of risk communication. First, it validates the interactive process of risk communication. Second, the definition recognizes that risk communication, by its nature, involves multiple and often competing messages. We discuss these two elements next.

Interactive Process

The most critical element of the National Research Council definition emphasizes that risk is an interactive process. The interaction occurs among all stakeholders in a risk setting. We define *stakeholders* as any persons or group of persons whose lives could be impacted by a given risk. Early views on risk communication paid little attention to stakeholder concerns or opinions. This limited view was based on a linear, unidirectional view of risk communication. Heath (1995) aptly captures the overly simplistic and biased consequence of viewing risk communication as the one-way dissemination of information. He explained that such a view perceives risk communication as "a linear, hypodermic communication process, whereby technical information can be injected into non-technical audiences" (p. 269). Williams and Olaniran (1998) also reject the linear view, arguing that risk communication cannot reach its potential to serve the public unless the communication exchange is viewed as "a dialogue instead of a monologue" (p. 393). Pursuing risk communication as an interactive process has the potential to make risk messages "increasingly effective and satisfying to the lay public who typically bear the risks" (Palenchar & Heath, 2002, p. 129).

Chess (2001) calls for interactive communication on a broad scale. She cautions organizations against perceiving the lay public as incapable of understanding risk messages. Instead, she advocates involving the lay public in decision-making, rather than seeking to protect them from whatever the organization deems dangerous. Witte (1995) insists that such interaction is essential for avoiding bias in organizational decision-making. She contends that no matter how considerate an organization tries to be in its linear risk communication, "there is no such thing as a neutral risk message" (p. 251). In response to this need for interactive risk communication, Heath et al. (2002) propose a current risk communication model that "places emphasis on dialogue, conflict resolution, consensus-building, and relationship development among the parties involved with or affected by the risk" (p. 318).

For risk communication to reach the objectives of dialogue, conflict resolution, and consensus building, government agencies and organizations must take into account the lay public's fears and frustrations, as well as any relevant technical information. In many instances, "the 'expert' view of what is likely to be important in making decisions about emerging technologies may not tally with the view of the public" (Frewer et al., 2003, p. 1131). Sandman (2000) identifies the emphasis on hazard over outrage as the recurrent source of such dispute. He explains that the *hazard* of a risk situation is based on technical analysis and risk calculation. *Outrage* represents the lay public's concern about a risk issue. For example, while the risk of contracting bovine spongiform encephalopathy (BSE), or mad cow disease, from eating United States beef is extremely low, public concern is very high. In fact, in 2003, when a single cow was diagnosed with BSE, nearly all United States beef exports were suspended by trading nations until confidence could be restored. Thus, to achieve the level of dialogue with the lay public, "the Golden Rule for risk managers is: always focus on the linked hazard-plus-concern" (Leiss, 2003, p. 369).

Risk messages fail miserably when they do not account for public concern. If, for example, government agencies hold forums simply to explain and justify decisions that have already been made, "rather than establishing two-way communication, some critics argue that holding a public meeting is the surest way for government agencies to minimize citizen input into decision making" (McComas, 2003, p. 165). Research suggests that, as public concern rises, the lay public is even less satisfied with such meetings. To establish public support through interactive communication, Heath and Palenchar (2000) advocate the blending of "community relations and risk communication" (p. 132). This type of blended effort can "build community support through collaborative, community-based decisions regarding the kinds of risks that exist and the emergency response measures that can be initiated as needed for public safety" (p. 132). Such community relations efforts establish a clear connection between government agencies or industry and their stakeholders.

Horlick-Jones et al. (2003) provide an insightful summary of government's role for establishing an interactive dialogue in risk communication. They contend that "government agencies in particular must be unflinchingly even-handed, and above all *independent* in all their risk communication activities" (p. 283). In so doing, these agencies play the role of "honest broker" of information the lay public needs to make informed risk issue decisions (p. 283). Even when government agencies and organizations willingly exchange information and invite participation, however, the risk communication process is further complicated by the tendency for multiple messages on a given risk topic to contradict each other.

Multiple Messages

Because risk communication is based on calculations and interpretations, most risk situations are replete with technical experts providing multiple messages that compete for acceptance. As we discussed earlier, the inherent uncertainty and ambiguity of risk situations makes such deliberation an inevitable part of risk communication. *Uncertainty* stems from a lack of information. Risk messages seek to alleviate this uncertainty by generating and assessing the credibility of available evidence. *Ambiguity* occurs when the available information is interpreted in more than one way and the quality or appropriate application of this evidence is debated. In responding to ambiguity, risk messages argue in favor of one interpretation of existing evidence over another. For example, mounting evidence suggests that the earth's temperature is rising. Yet the interpretation of this evidence results in intense public debate. Many environmentalists, claiming that fossil fuels are the cause of global warming, call for a shift to cleaner sources of energy. Conversely, their opponents argue that the environmentalists' claim that the earth's temperature is rising is based on uncertain evidence. In addition, those who argue against the recommendations of environmental groups insist that, with feasible modifications, fossil fuels such as coal can be a "clean" source of energy. Whichever the stand a person takes on the issue of global warming, however, the fact is that there are multiple messages attending to the uncertainty and ambiguity of the situation.

Debates that pit risk messages against one another are fostered, to a large extent, by competing interests among organizations and government agencies. Naturally, in a risk debate one interpretation of a risk situation will favor some participants over others. Palenchar and Heath (2002) explain that, as a consequence of this debate, competing parties in risk situations often confuse the lay public with "competing scientific conclusions" (p. 130) that portray the "same situation from different symbolic realities" (p. 136). Rather than reducing uncertainty, these "debates among experts are apt to heighten public uncertainty about what the facts really are, increase doubts about whether the hazards are really understood, and decrease the credibility of official spokespersons" (Kasperson et al., 2000, p. 242). For example, higher air quality standards may benefit the physical health of residents living near a factory. Conversely, imposing higher standards may be costly for the factory, thereby diminishing its profits and causing a loss of jobs in the area. The debate, then, becomes one of how high to raise air quality standards while considering the constraint of financial cost. This complexity of interests is common in risk communication. Making the situation even more complex is, "the fact that, in the US, risk-management tends to rely heavily upon an adversarial legal system that pits expert against expert, with each expert contradicting the other's risk assessments and further destroying the public trust" (Slovic, 2000, p. xxxv). Regardless of how the lay public assesses competing messages, Scherer and Juanillo (2003) explain that, "what is important is not to lose sight of the end goal, which is to provide an arena for participants to gain insight, reflect on the issues, and provide better informed recommendations or decisions" (p. 234).

Construing Risk Messages

The lay public must construe relevance and meaning from a given risk issue's myriad messages and relationships. By *construe*, we mean the public must infer meaning by assessing the importance and accuracy of the information and the authenticity of the source. As Fig. 1.2 indicates, public audiences engage in this process of interpretation on multiple levels.

In an effort to understand competing claims in risk situations, the general public may evaluate the messages systematically or heuristically (Kahlor et al., 2003, p. 356). A *systematic evaluation* involves "cognitive effort" (p. 365) to consider the evidence of the messages and determine the accuracy of the competing claims. Conversely, when audiences interpret the competing messages *heuristically* they take a more passive role in the interpretation of evidence. Instead, the lay public focuses on the "source's identity or other non-content cues" (p. 365).

The complexity of technological advancements makes the detailed comprehension of all logical aspects of each risk issue impossible. Despite this lack of expertise, the lay public possesses a "sense of risk" that "arises from the normal concerns about the probability that a risk could occur that causes varying amounts of damage to people or the environment" (Palenchar et al., 2005, p. 61). This sense of risk compels the public to attend to or seek out information related to their perceived sense of vulnerability. This process of information collection enables the public to make "risk judgment[s]" based on their "perceived probability" that a risk will occur and its "perceived severity" (Griffin et al., 2004, p. 29). *Perceived probability* refers to the likelihood that a risk situation will evolve into a crisis event. *Perceived severity* addresses the seriousness or level of harm that could occur in the case of a crisis. An event with low probability and low severity would instill little concern. In

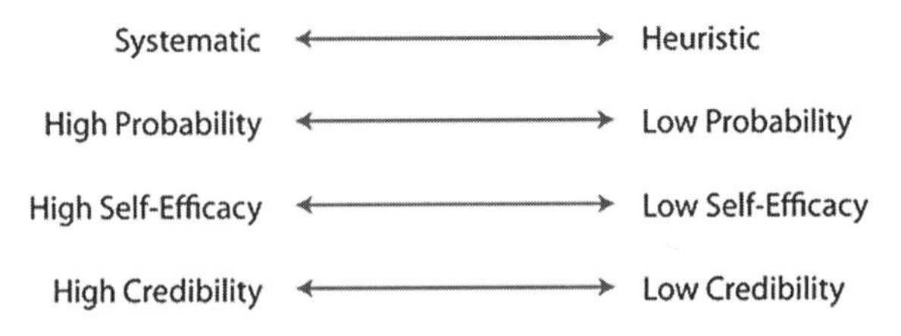

Fig. 1.2 Approaches for construing risk communication messages

contrast, an event with high probability and high severity would warrant immediate action.

Self-efficacy is a third vital element in the public's sense of risk. Self-efficacy "is the conviction that one can successfully execute the behavior required to produce outcomes" (Gordon, 2003, p. 1287). Generally speaking, individuals who have a strong sense of risk about a given subject are more likely to respond to messages they believe provide a reasonable strategy for personally reducing their level of risk. If, however, individuals "entertain serious doubts about whether they can perform the necessary activities such information does not influence their behavior" (Gordon, 2003, p. 1287). Similarly, the public's ability to make informed decisions regarding self-efficacy is "substantially increased when resources are made available and under their control;" however, "access to information without an enhanced capacity for action will only frustrate individuals seeking to acquire more information" (Heath et al., 2002, p. 323). Ideally, risk messages provide some level of self-efficacy for stakeholders. Without self-efficacy, the perceived relevance of risk communication is narrowed.

Transcending all systems of meaning is the importance of trust and credibility. Key elements of *credibility* include "openness, accuracy, trustworthiness, impartiality, and completeness of information provided to citizens" (McComas, 2003, p. 180). Similarly, *trust* is based on the degree to which a person or organization "is competent, objective, fair, consistent, having no hidden agenda, and being genuinely concerned about the vulnerability of its stakeholders" (Palenchar et al., 2005, p. 62). The failure to establish trust makes risk communication futile. Slovic (2000) purports that "the limited effectiveness of risk-communication efforts can be attributed to the lack of trust. If you trust the risk manager, communication is relatively easy. If trust is lacking, no form or process of communication will be satisfactory" (p. xxxv). To proactively avoid the disabling consequence of lost trust, Heath and Palenchar (2000) emphasize that, "savvy companies seek to build relationships with members of adjacent communities on the assumption that they will enjoy support rather than oppositions" (p. 136). Efforts to build such relationships will fail, however, unless they are genuine.

Meaningful Access

When we use the phrase *meaningful access*, we are referring to both opportunities for interaction with key decision-makers and for acquiring the information necessary to make informed judgments about a risk issue. We begin our discussion of meaningful access by explaining the role *frame of reference* plays in the interpretation of risk messages. We then introduce the concept of *entente* as an indispensable prerequisite for meaningful access.

The simple exposure to information, however, does not translate to understanding. Heath et al. (2002) argue that "without understanding, information is not knowledge" (p. 325). Scholars have taken a variety of approaches in their efforts to understand how the lay public interprets, understands, and responds to risk messages. The psychometric paradigm focused largely on developing cognitive maps for measuring the attitudes and perceptions of individuals (Slovic, 1987). This research has evolved to include, "underlying social and cultural values and contexts associated with risk perceptions," focusing on "predominant worldviews or cultural biases in which the individual participates or identifies with such as hierarchy, fatalism, individualism and egalitarianism" (Scherer & Juanillo, 2003, p. 227). Drawing on the work of Kenneth Burke, Hamilton (2003) characterizes this expanded view as the public's frame of acceptance. A *frame of acceptance* refers to an "individual's orientation that is drawn from and combines elements of larger meaning systems or orientations" (p. 292). Thus, segments of the population interpret messages distinctly based on "diverse systems of meaning" that are based on their frame of reference (Hamilton, 2003, p. 301).

Poumadere and Mays (2003) recognize that "the conflicts around risk issues are above all, for the players involved, composed of disturbing feelings of gap between their vision and that which they believe to be the vision of others" (p. 235). Emphasizing one view over another as "the 'real' risk at hand, or the priority that should be given to risk reduction" rebuffs the frame of reference held by any group that holds a different view of the situation (p. 235). As such, any dissenting group would be denied access to the risk communication dialogue. For meaningful access to occur, Poumadere and Mays contend that "a shared sense of reality is a necessary prerequisite to communication inside or across any of these bounds" (p. 235). They label this shared sense of reality, *entente*. Likening entente to oxygen, Poumadere and Mays explain that "entente's existence and role only becomes apparent when it comes to be in short supply" (p. 235). They explain that, "in the domain of risk, any player can deprive another or others of entente by imposing, intentionally or not, a vision of reality that contradicts the others' beliefs and sense of reality" (p. 236). This is not to say that two parties cannot disagree in a risk debate. Rather, both sides can and should begin the debate by recognizing and respecting the opinions of those who hold a different frame of reference. When entente is a part of a risk debate, "science and the irrational are not seen to be mutually exclusive: they can exist side by side in the construction of a risk event" (p. 240). By acknowledging distinct views, the concerns of all participants in the debate are considered. Specifically, "pre-existing or emergent concerns, doubts, worries about various objects in the environment are not dismissed, but taken on board as the starting point for expertise and evaluation" (p. 240). The ultimate goal of a risk discussion with entente is to provide a framework, formal or informal, that can "accommodate the expression of different worldviews" (p. 240).

A Message-Centered Focus

The interactive nature of risk communication, in tandem with the multiple, often conflicting messages on any given risk issue, leads us to view risk communication as a process of interacting arguments. This perspective differs notably from other approaches to risk communication. Most importantly, by viewing risk communication from the perspective of interacting arguments, we see *messages* as our primary focus. By contrast, economic, sociological, and psychological perspectives of risk place their primary emphasis elsewhere.

Economic models of risk, for instance, seek to understand the *financial cost* of containing a risk. Such models tend to consider the likelihood and cost of a crisis. If the likelihood is high and the cost is high, an economic calculation would recommend devoting considerable resources to limiting the risk. For example, auto manufacturers now include airbags with all automobiles because the potential reward, safety, outweighs the financial cost. On the other hand, if the potential toll or the likelihood of a crisis event is low, an economic calculation would suggest devoting fewer resources to reducing that risk. From an economic perspective, risk communication is focused largely on policy. Calculations are completed and shared in order to determine how resources will or will not be expended on a given risk issue.

Conversely, sociological approaches to risk focus on the *behavior trends* in the general population. Sociologists seek to understand how members of a society recognize and respond to risks. Why, for example, do people choose to smoke even though a preponderance of evidence indicates that smoking is bad for their health? Why does obesity continue to rise in spite of efforts to educate the general public on the merits of exercise and a healthy diet. From a sociological perspective, risk communication is focused primarily on communication campaigns. Risk issues are identified, recommendations are formed, and the recommendations are then shared through elaborate communication campaigns. These campaigns are then evaluated from a broad perspective.

Finally, psychological approaches to risk communication seek to uncover generalizable trends in individual behavior. Some psychological research traces brain activity in hopes of understanding decision-making in risky situations. Other psychological research seeks to understand how individuals are influenced by others in situations involving risk. In general terms, the psychological perspective sees risk communication as a stimulus that produces a response in the individual. By understanding the potential of various stimuli, psychologists can make recommendations for influencing behavior.

While each of these perspectives has merit, we believe our specific focus on the interpretation of competing messages through an interactive process provides insights that complement the others.

Risk Communication as Interacting Arguments

Our interacting arguments perspective is based on the work of Perelman and Olbrechts-Tyteca (1969). They explain that opposing arguments on a given issue are best understood systematically because the argumentative situation "shifts each moment as argumentation proceeds" (p. 460). They explain that the argumentative situation advances in two ways:

(1) "[B]y a more thorough, closer, or differently conducted analysis of the statements made"
(2) "[B]y giving consideration to an increasing number of spontaneous arguments having the discourse as their subject." (p. 460)

Thus, to understand arguments associated with a given risk issue, we must understand both the technical information and the public's understanding of or discourse about the risk issue.

As the arguments interact, the strength and weakness of the claims are assessed by those offering formal arguments and by those who engage in discourse about the issue. In other words, opponents debate the accuracy of claims made by their adversaries while the public engages in a less formal dialogue about who to believe. When two opposing sides of an issue offer conflicting arguments; however, there is rarely a complete distinction between them. Thus, observers can rarely conclude that one party is completely right and the other party is completely wrong. Rather, as arguments interact in the system of discourse, there are typically some degrees of *convergence*. Perelman and Olbrechts-Tyteca (1969) explain that convergence occurs when "several distinct arguments lead to a single conclusion" (p. 471). The "strength" of converging arguments is "almost always recognized" because the "likelihood that several entirely erroneous arguments would reach the same result is very small" (p. 471).

We see convergence as the primary objective in risk communication. The uncertainty in risk situations gives rise to competing claims about the levels of danger and about the appropriate means for responding. Thus, diverse arguments emerge. As the public observes these arguments, it is unlikely to fully accept one line of reasoning and totally reject another. Instead, the public is likely to *make sense* of the issue by observing ways in which the arguments interact. As Fig. 1.3 indicates, convergence occurs when distinct bodies of knowledge overlap, resulting in some capacity of agreement. When considering the risk of genetically modified foods, for

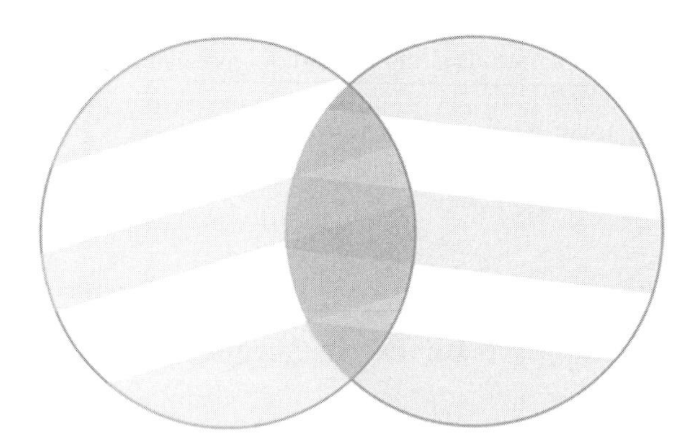

Fig. 1.3 A basic view of convergence

example, opponents argue that such modifications could wreak havoc on the ecosystem. Proponents of genetically modified foods recognize the potential for such harm to occur, but claim that provisions can be made to contain the fields where genetically modified crops are grown. Although the two groups have strongly opposing viewpoints, the interaction of their arguments produces convergence on the potential of environmental risk. With this convergence in mind, observers can begin to form an opinion about how safe is safe enough when growing genetically modified crops.

The perspective of interacting arguments enhances our understanding of risk communication in two ways. First, those who take opposing sides in risk situations engage in argument. As in a debate, they take conflicting stands and offer justification for their positions. Yet if we view risk communication as a debate, we suggest that there is a winner and a loser. This is rarely the case. Typically, each side has some level of merit in its claims. Realistically, arguments interact to a point where some convergence is reached. This convergence, then, is a major step toward resolution.

Second, the perspective of interacting arguments suggests that the discourse on risk issues operates systematically. Observers do not sit idly as they await the next formal argument or rebuttal from a technical expert on a given risk issue. Rather, observers collect and contemplate information from a variety of sources – some credible, others highly biased – and discuss this information with their family, friends, and neighbors. These discussions lead to a variety of opinions. Approaching risk communication, therefore, from the perspective of interacting arguments not only enhances our understanding of risk communication, it also accounts for the complexity of the risk communication process.

In the next section we elaborate on convergence's distinction from congruence, mutual exclusivity, and dominance.

Convergence Versus Congruence

Initially, we must distinguish convergence from congruence. *Congruence* implies that all parties have settled on a single, unifying interpretation of a risk situation. In rare instances, a dialogue may lead to consensus, but in most cases debate over risk issues lingers continually over some controversial aspects. Chess and Clarke (2007), for example, argue for transparency among agencies in the discussion of risk issues. The objective of such transparency is to expose rather than conceal conflicting evidence. Thus, risk communication is as likely to expose differences as it is to create single-mindedness. Figure 1.4 portrays congruence. Each circle represents the body of arguments on two opposing sides; the third circle represents the complete agreement achieved between the two sides. The complex nature of risk communication makes such congruence highly unlikely. This is not to say that working toward congruence is unwise. Rather, we see congruence as a destination seldom reached in the journey that constitutes a risk communication dialogue on a given issue.

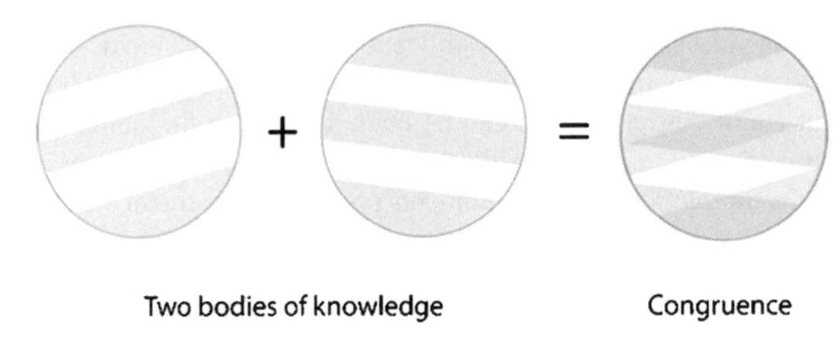

Fig. 1.4 Distinguishing convergence from congruence

Convergence Versus Mutual Exclusivity

Although risk communication is frequently referred to as a public debate, as noted, there is no clear winner and loser. Instead, conflicts over competing risk messages create a form of dialogue that persists over an extended period of time. Murdock et al. (2003) explain that the perception of mutual exclusivity in risk communication is "rooted in the classical opposition between 'them' and 'us' at the heart of populism" (p. 176). This mentality typically established the lay public as an antagonist to government or industry. We find the perception that bodies of risk information are mutually exclusive overly simplistic. Discarding or discrediting one body of knowledge for another has the potential to diminish the overall understanding of a risk debate.

In Fig. 1.5, we distinguish between convergence and mutual exclusivity. Convergence occurs when there is some degree of agreement between two bodies of argument on seemingly opposing sides. If the arguments are mutually exclusive, one entire body of argument is discredited. We find a *winner take all* mentality in risk communication imprudent for two reasons. First, from this perspective the competitive nature of risk communication prohibits any form of synergy among the claims. Second, whenever one body of knowledge is muted, the lay public has been denied access to that information. Hence, a critical element of the risk communication process is lost.

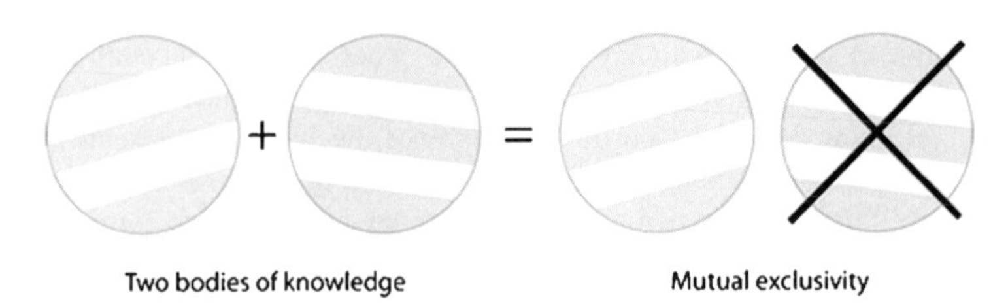

Fig. 1.5 Distinguishing convergence from mutual exclusivity

Convergence Versus Dominance

Throughout this chapter, we have emphasized the importance of recognizing that an interaction among parties with differing views on a risk situation is essential to effective risk communication. Without a respect for differing world views, the quality of the risk communication is diminished greatly. Thus, any effort to deny the lay public access to any form of argument is detrimental to the risk communication process. If an industry or government agency imposes a dominating presence in the public discussion of a risk issue, entente is lost. As Poumadere and Mays (2003) argue, risk communication "should be open enough to accommodate the expression of different worldviews" (p. 240). No party has the right to dominate and exclude the concerns of another, "not even those who will ultimately be proved right by events" (p. 240). As is depicted in Figure 1.6, dominating communication in a risk event mutes the opposition, thereby denying the lay public the opportunity to consider any potential points of convergence between the two bodies of knowledge. Whenever the opportunity for convergence is lost, there is an increased likelihood that the quality of the decisions ultimately emerging from the discussion will be diminished (Fig. 1.6).

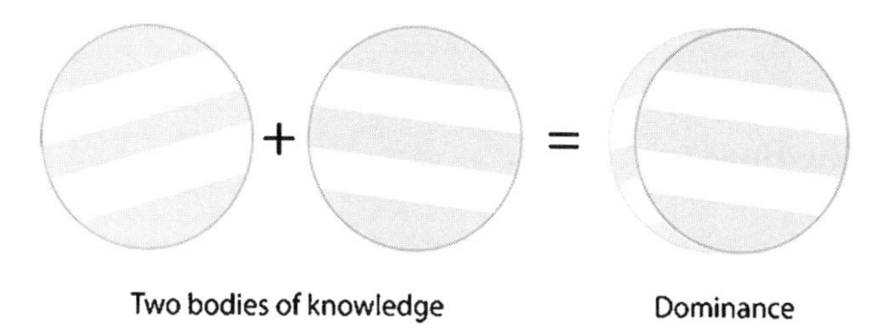

Fig. 1.6 Distinguishing convergence from dominance

Convergence and Multiple Sources

For purposes of clarity, we portrayed convergence in the previous diagram with only two opposing bodies of knowledge. Realistically, there are many bodies of knowledge contributing to risk communication on any given topic. Government agencies introduce proposals for regulation. Technical experts approach a risk issue with divergent points of view and conflicting evidence. Special interest groups establish claims around a risk issue in an effort to protect their supporters. Individual politicians or opposing political parties can introduce claims that contrast with the status quo in government policy. Individuals have interpersonal networks from which they derive information and opinions. We view all of these separate bodies of knowledge and subsystems within the larger system of discourse surrounding a risk issue. As Fig. 1.7 indicates, convergence is complex and multi-faceted

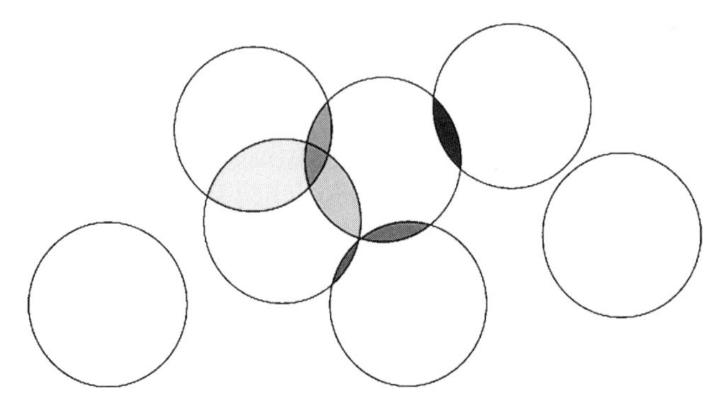

Fig. 1.7 Convergence on a systematic level

when viewed from this systematic perspective. Nevertheless, viewing risk messages from the perspective of convergence provides a means for better understanding how arguments interact to form the knowledge base upon which the lay public forms opinions.

Focus of the Book

In this book, we offer examples of risk from many aspects. One of our primary areas of focus, however, is on the food industry. We observe that the increasing demand for processed and prepared food, along with environmental challenges and increased reliance on technology to provide the world's food supply, have created increased uncertainty and risk in the area of food safety and protection. For example, "In the United States, foodborne diseases are estimated to cause 6 million to 81 million illnesses and up to 9,000 deaths each year" (Mead et al., 1999, p. 607). Complicating the food industry further is the threat of bioterrorism. Following 9/11, American troops made a troubling discovery in Afghanistan. As the Taliban and al Qa'eda forces retreated from their caves and safe houses, American troops "found hundreds of pages of U.S. agricultural documents that had been translated into Arabic" (Doeg, 2005, p. 167). Most alarming, a "terrorist's training manual was reportedly devoted to agricultural terrorism such as the destruction of crops, livestock, and food processing operations" (pp. 167–168). With these concerns in mind, we see the food industry as a viable area for the exploration of effective risk communication.

Summary

The inherent uncertainty in our lives makes risk an inevitable factor in nearly every decision we make. Thus, as the National Research Council noted, risk communication is a fundamental part of a democratic society. Any time risk communication is reduced to a linear process, the efficacy of the democratic ideal is threatened. To reach its potential, risk communication must consist of an interactive process where all parties are given access to multiple messages representing all relevant worldviews. Identifying points of convergence serves as a means for making sense of these interacting arguments. Convergence is distinct from congruence in that convergence does not assume complete agreement. Because elements of distinct bodies of knowledge may coincide, convergence discredits the assumption that competing risk arguments are mutually exclusive. Because access to all worldviews is an essential element of the interacting process, imposing dominance of one view over another is unacceptable in effective risk communication. Finally, arguments can interact on a systematic level. A single observer can find points of convergence on many levels from many sources. Individuals consider these multiple points of convergence as they construe meaning and make decisions about a given risk situation.

References

Beck, U. (1992). *Risk society: Towards a new modernity*. Thousand Oaks, CA: Sage.

Chess, C. (2001). Organizational theory and stages of risk communication. *Risk Analysis, 21*, 179–188.

Chess, C., & Clarke, L. (2007, September). Facilitation of risk communication during the anthrax attacks of 2001: The organizational backstory. *American Journal of Public Health, 97*, 1578–1583.

Doeg. C. (2005). *Crisis management in the food and drinks industry: A practical approach* (2nd ed.). New York: Springer-Science + Media, Inc.

Frewer, L. J., Scholderer, J., & Bredahl, L. (2003). Communicating about the risks and benefits of genetically modified foods: The mediating role of trust. *Risk Analysis, 23*, 1117–1133.

Gordon, J. (2003). Risk communication and foodborne illness: Message sponsorship and attempts to stimulate perceptions of risk. *Risk Analysis, 23*, 1287–1296.

Griffin, R. J., Neuwirth, K., Dunwoody, S., & Giese, J. (2004). Information sufficiency and risk communication. *Media Psychology, 6*, 23–61.

Hamilton, J. D. (2003). Exploring technical and cultural appeals in strategic risk communication: The Fernald radium case. *Risk Analysis, 23*, 291–301.

Heath, R. L. (1995). Corporate environmental risk communication: Cases and practices along the Texas Gulf Coast. In B. R. Burelson (Ed.), *Communication Yearbook 18* (pp. 255–277). Thousand Oaks, CA: Sage.

Heath, R. L., Bradshaw, J., & Lee, J. (2002). Community relationship building: Local leadership in the risk communication infrastructure. *Journal of Public Relations Research, 14*, 317–353.

Heath, R. L., & Palenchar, M. (2000). Community relations and risk communication: A longitudinal study of the impact of emergency response messages. *Journal of Public Relations Research, 12*(2), 131–161.

Horlick-Jones, T., Sime, J., & Pidgeon, N. (2003). The social dynamics of environmental risk perception: Implication for risk communication research and practice. In N. Pidgeon, R. E. Kasperson, & P. Slovic (Eds.), *The social amplification of risk* (pp. 262–285). Cambridge, United Kingdom: Cambridge University Press.

Kahlor, L., Dunwoody, S., Griffin, R. J., Neuwirth, K., & Giese, J. (2003). Studying heuristic-systematic processing of risk communication. *Risk Analysis, 23*, 355–367.

Kasperson, R. E., Ortwin, R., Slovic, P., Brown, H. S., Emel, J., Goble, R., Kasperson, J. X., & Ratick, S. (2000). The social amplification of risk: A social framework. In P. Slovic, (Ed.), *The perception of risk* (pp. 232–245). London: Earthscan Publications Ltd.

Kasperson, J. X., Kasperson, R. E., Pidgeon, N., & Slovic, P. (2003). The social amplification of risk: Assessing fifteen years of research and theory. In N. Pidgeon, R. E. Kasperson, & P. Slovic (Eds.), *The social amplification of risk* (p. 13–46). Cambridge, United Kingdom: Cambridge University Press.

Leiss, W. (2003). Searching for the public policy relevance of the risk amplification framework. In N. Pidgeon, R. E. Kasperson, & P. Slovic (Eds.), *The social amplification of risk* (pp. 355–373). Cambridge, United Kingdom: Cambridge University Press.

McComas, K. A. (2003). Citizen satisfaction with public meetings used for risk communication. *Journal of Applied Communication Research, 31*, 164–184.

Mead, P. S., Slutsker, L., Dietz, V., McCaig, L. F., Bresee, J. S., Shapiro, C., Griffin, P. M., & Tauxe, R. V. (1999). Food-related illness and death in the United States. *Emerging Infectious Diseases, 5*, 607–625.

Mitroff, I. I., & Anagnos, G. (2001). *Managing crises before they happen: What every executive and manager needs to know about crisis management.* New York: AMACOM.

Murdock, G., Pets, J., & Horlick-Jones, T. (2003). After amplification: Rethinking the role of media in risk communication. In N. Pidgeon, R. E. Kasperson, & P. Slovic (Eds.) *The social amplification of risk* (pp. 262–285). Cambridge, United Kingdom: Cambridge University Press.

National Research Council (1989). *Improving risk communication.* Washington, D.C.: National Academy Press.

Palenchar, M. J., & Heath, R. L. (2002). Another part of the risk communication model: Analysis of communication processes and message content. *Journal of Public Relations Research, 14*(2), 127–158.

Palenchar, M. J., Heath, R. L., & Orberton, E. M. (2005). Terrorism and industrial chemical production: A new era of risk communication. *Communication Research Reports, 22*(1), 59–67.

Perelman, C., & Olbrechts-Tyteca, L. (1969). *The new rhetoric: A treatise on argumentation.* London: University of Notre Dame Press.

Poumadere, M., & Mays, C. (2003). The dynamics of risk amplification and attenuation in context: A French case study. In N. Pidgeon, R. E. Kasperson, & P. Slovic (Eds.), *The social amplification of risk* (pp. 209–242). Cambridge, United Kingdom: Cambridge University Press.

Sandman, P. (2000). Open communication. In E. Mather, P. Stewart, & T. Ten Eyck (Eds.), *Risk communication in food safety: Motivating and building trust.* East Lansing: National Food Safety and Toxicology Center, Michigan State University.

Scherer, C. W., & Juanillo, N. K. (2003). The continuing challenge of community health risk management and communication. In T. L. Thompson, A. Dorsey, K. Miller, & R. Parrott (Eds.), *Handbook of health communication* (pp. 221–239). Mahwah, NJ: Lawrence Erlbaum.

Seeger, M. W., Sellnow, T. L., & Ulmer, R. R. (2003). *Communication and organizational crisis.* Westport, CT: Praeger.

Slovic, P. (1987). Perception of risk. *Science, 236*(4799), 280–285.

Slovic, P. (2000). Introduction and overview. In P. Slovic, (Ed.), *The perception of risk* (pp. xxi–xxxvii). London: Earthscan Publications Ltd.

Williams, D. E., & Olaniran, B. A. (1998). Expanding the crisis planning function: Introducing elements of risk communication to crisis communication practice. *Public Relations Review, 24*, 387–400.

Witte, K. (1995). Generating effective risk messages: How scary should your risk communication be? In B. R. Burelson (Ed.), *Communication Yearbook 18* (pp. 229–254). Thousand Oaks, CA: Sage.

Chapter 2
Best Practices for Risk Communication

It doesn't work to leap a twenty-foot chasm in two ten-foot jumps. (American Proverb)

In Chapter 1, we reviewed the major issues in risk communication and outlined a view of the public communication process that often accompanies discussions of risks. Often, these discussions digress into arguments over entrenched positions with little hope of achieving consensus about what is an acceptable risk. Trust is undermined, the public does not have access to information, and antagonistic relationships ensue. We suggested that even in these cases, there is an opportunity to approach risk communication by recognizing that positions can converge and a middle ground in discussions of risk can be created. We view this convergence as an opportunity for dialogue leading to a more effective form of risk communication.

Another aspect to this convergence approach involves grounding activities in best practices.[1] The best practices method has been used widely in organizational and professional settings as a way to identify and institutionalize a set of tested industry practices. The success of these practices is considered best in the sense that they are more likely to achieve desired outcomes. Best practices also help establish consistency across a set of related activities and contexts. In this way, industry norms may develop. In addition, a best practices approach to risk communication allows for a broader grounding in larger principles of effective communication and a longer-term perspective.

In this chapter, we first describe the best practices approach. This includes the way it has been used in other communication contexts. Next, we describe nine best practices of risk communication that improve effectiveness by building constructive relationships with risk stakeholders, acknowledging the complex and multi-dimensional nature of both risk and communication and responding to the communication and informational needs of diverse audiences.

Concept of Best Practices

The best practices method is a popular management approach to improving organizational and professional practice in a wide array of venues, including diverse

T. L. Sellnow et al. *Effective Risk Communication*

DOI: 10.1007/978-0-387-79727-4_2

communication contexts (Seeger, 2006). In many professional contexts, best practices become the industry standard, used to establish a basic framework for a set of similar activities. When these industry standards emerge, more information is available about their implementation and both operators and regulatory agencies have a clearer understanding of expectations. The best practices method has been used widely in manufacturing settings, including in food production. Best practices also have been used in such diverse areas as corporate communication, health communication, training and development, public relations, employee communication, stakeholder communication, and the communication of change, among many others.

The best practices approach is a form of grounded theory for process and organizational improvement. Moreover, they have the advantage of being tested in industry contexts. The identification of best practices is often associated with benchmarking and larger process and quality improvement initiatives and programs of strategic organizational change (Kyro, 2004). Process improvement as a tool of management generally involves a systematic overview, analysis, and assessment of organizational processes in an effort to improve quality and efficiency. Often, a benchmarking process of identification of industry standards through a focus on industry leaders and recognized experts in a given field is the first step. Benchmarking reviews often proceed with systemic observation, description, and measurement of what are acknowledged as high quality and efficient organizations (Ahmed & Rafiq, 1998). The processes, practices, and systems identified among industry leaders are then described as *best practices*. These can provide useful models for other organizations with similar functions, contingencies, and missions. In addition, panels of experts in a given field are sometimes asked to generate specific normative standards and principles characteristic of effectiveness and efficiency (Seeger, 2006). Best practices, then, usually take the form of a general set of standards, procedures, guidelines, norms, reference points, or principles that inform practice and are designed to improve performance in specific ways.

Best practices are generally practice-driven but may also be anchored in systemic research and a grounded theoretical approach. Grounded theory is particularly useful in developing generalized standards and principles as an area of organizational and professional practice matures and develops. Grounded theory proceeds from an inductive standpoint and seeks to describe patterns and conceptual categories within a particular data set (Glaser & Strauss, 1967). This inductive approach has been used widely in communication inquiry (Gilchrist & Browning, 1981; Nicotera, 1983). Best practice categories can then be explored in other contexts to determine if they can be generalized. An essential consideration, however, is that professional and organizational contexts are diverse, dynamic, and complex. What works in one industry may have very limited applicability to another. Thus, widespread adaptation of best practices should be undertaken cautiously with a firm understanding of contextual factors and situational variables.

These approaches to the improvement of professional and organizational processes and practices may also be framed within the larger context of organizational learning. Organizational learning concerns the ways by which organizations acquire new information and knowledge, retain information, and translate that information into skills and practices. Best practices are useful ways to package learned principles and skills in ways that facilitate their communication both within and between organizations and ultimately their widespread adoption (Cohen & Sproull, 1996). Thus, best practices can also be understood as larger lessons for organizational and professional learning for a particular venue of practice.

Best Practices for Risk Communication

We describe nine best practices for risk communication that are generated from the research literature on risk communication and from the larger goal of achieving convergence around issues of risk. The view of best practices presented here also is informed by fundamental values of communication including openness, honesty, equity, and fairness. Moreover, these best practices are designed to help build constructive and mutually beneficial relationships with risk stakeholders, acknowledge the complex and multi-dimensional nature of both risk and communication, and respond to the communication and informational needs of diverse and changing audiences.

Infuse Risk Communication into Policy Decisions

Issues and questions of risk usually must be addressed in an organization's or agency's larger policy position. Organizations often have a vested interest in a particular interpretation of a risk (Morgan et al., 2001). Policy establishes a formal organizational position regarding some issues so that subsequent decisions are consistent and reflect the overall interests of the organization or agency. Policies about risk may evolve and be communicated in a variety of ways. Often, an organization's policy position is inherited from tradition and past practices. In cases where risks have developed incrementally over time, as is the case with many environmental risks, organizations typically replicate previous policy positions and interpretations, perhaps with incremental changes. Communication in these cases usually becomes the method for explaining or justifying the pre-existing policy and can be expected to reiterate previous arguments and positions. In other cases, the communication process itself creates policy. Public statements about risk made by senior officials or top managers are often widely distributed in the press. In these cases, the policy may emerge to support or be consistent with these statements. A third method for policy development involves a systemic and comprehensive review where policy is informed and developed by information from a variety of sources. These reviews, however, are typically dominated by empirically driven risk calculations and estimations. In these cases, communication may be part of this review, or may simply be brought in after the fact to disseminate the policy; i.e., sell the decision after it has been made.

While the relationship between communication and policy formation may take several forms, it is most effective when communication functions cooperatively as part of the policy formation process (Dozier & Broom, 1995). Setting the overall risk policy for an organization requires the input of both the fields of risk estimation and risk communication.

Treat Risk Communication as a Process

Most contemporary understandings of communication emphasize its process features. When communication is viewed as a process, its dynamic, interactive and adaptive elements are placed in the foreground (Berlo, 1977). Communication, accordingly, is influenced by contextual dynamics, sender and receiver features, message attributes, and elements of "noise" or anything that might interfere with shared meaning. These elements are constantly changing in both substances and in their relationship to one another. What was said previously influences subsequent messages and interpretations. In contrast, many early views of communication adopted a hypodermic needle model, which viewed the process as static and the receiver as passive. This model assumes that a message can be injected into a receiver and somehow the desired outcomes will occur. This assumption has lead to many dramatic failures in communication. Similarly, many organizations treat risk communication as product and assume that once the message has been produced and delivered to an audience, the desired outcome is by definition achieved (Kasperson, 1991).

One of the most important elements of a process view of communication is feedback. Feedback is an adaptive element to the communication system allowing messages to be refined and improving the probability for success. For example, a receiver might respond to a statement by saying, "I don't understand." The sender can then adjust the message to help the receiver understand. Most advertising messages, public relations campaigns, and many risk messages are tested with focus groups who provide feedback so that the message can be improved. Feedback is often described as a form of meta-communication that informs the participants about the communication process. By so doing, feedback allows for strategic adjustments to messages, channels, audiences, and contexts so that effectiveness can be improved.

Risk is similarly a dynamic process. Risks are not static features but change in and of themselves. Risk and uncertainty are closely associated. Moreover, both public and scientific understanding of risk also changes over time. As Heath et al. (1998) note, "discussions of risk refer to (a) the likelihood that some harmful event will occur and (b) the chance that it will harm the person making the risk attributions" (p. 42). Changes may occur in the predicted likelihood of harm and the changes of harm to a person.

Risk communication is most effective when it embraces these dynamic features. When communication and risk are treated as static, changes in risk factors are not communicated and the risk messages are viewed as one-time products and are not adapted to meet changing audiences and conditions as well as evolving understandings of risk.

Account for the Uncertainty Inherent in Risk

Risk is a dynamic phenomenon where changes in cultural and scientific understanding of risk are not static. New information about risks continually emerge. Risk factors interact with other variables in unanticipated, non-linear, and chaotic ways. No given risk assessment, no matter how comprehensive, can account for what is yet to be learned. Some researchers have described the inherent uncertainty in risk assessments with the concept of known unknowns and unknown unknowns (Conrow, 2003). Risk estimations may be able to recognize that some questions are unknown, and thus account for these known unknowns, but it is not possible to account for those factors that are unknown unknowns. Given the inherently dynamic and uncertain nature of risk, messages are most accurate and effective when they are stated in equivocal terms.

Remaining equivocal in risk messages means acknowledging that uncertainty exists and framing messages within that inherent uncertainty. Messages that include statements such as, "We do not yet have all the facts," and "Our understanding of these factors is always improving," can be used to preface risk messages. Some risk messages offer general advice about risk factors, such as food labels that include generalized food handling instructions which may help reduce risks in a variety of situations.

There are, however, often significant pressures to frame risk messages in absolute terms. In cases where the larger goal is to simply remove the risk question from the public agenda, risk communicators often fall back on overly reassuring messages. Statements such as, "There is no risk," "This item is perfectly safe," or "Any concern is irrational," are frequently communicated in these circumstances. These overly reassuring messages have the impact of undermining trust and compromising credibility. Even lay persons recognize that overly-certain projections of risks that fail to acknowledge the inherent uncertainty are simply unrealistic. They beg the question, how can anyone know for complete certainty how a risk might develop in the future? These overly-certain and reassuring messages also imply that the communicator is not being entirely open and honest regarding the nature of the risk.

Some communicators, particularly in public health, have become very sophisticated in offering risk messages that acknowledge the inherent uncertainty. In particular, top public health officials often begin risk messages by acknowledging, "We do not yet have all the facts." While such equivocal messages are easier to offer in circumstances where a risk is evolving and information is being collected, risk messages function most effectively when they avoid overly-certain predictions.

Design Risk Messages to be Culturally Sensitive

As described earlier, perceptions of risk are socially and culturally constructed and can be expected to vary widely based on several factors. In addition, specific features of the audience influence the way messages are received and interpreted. These features include gender, education, age, and culture.

Research, for example, has demonstrated that women are more receptive to risk messages than are men. Women also appear to respond to some kinds of risks with more problem solving strategies than do men. The ability to understand specific types of risk messages may be associated with educational level. There is also evidence that higher education is associated with greater use of Internet information sources. The elderly may be more susceptible to some risks due to underlying health factors. At the same time, they have broader frames of reference regarding risk and more experience in risk management. In general, younger people are more likely to take risks than are older people.

The increasingly diverse U. S. population reveals the need to view culture as an audience variable in risk communication. Some cultures may be risk adverse while others are more willing to take risks. Cultural preferences may influence the effectiveness of risk messages. In some cases, risk messages are most effective when they are delivered by a respected member of that particular community. These cultural agents generally have more credibility with the intended audience and understand cultural nuances. Messages delivered by outsiders are likely to be ineffective and may even create mistrust (Littlefield et al., 2006). Language differences are also critical audience variables in effective communication. Delivering risk messages in a culturally-sensitive and effective way requires more than simple translation.

Understanding the personal, community and cultural influences on risk perception enables communicators to tailor their communication strategies to audience characteristics and increase the probability of success. One strategy involves adapting the location and form of messages to fit the preference and media consumption patterns of the target audience. Risk messages about food, for example, may be most effective when they are distributed in media sources frequented by grocery shoppers. Most studies suggest that the general public's proficiency is at the eighth grade level or lower. Risk messages, therefore, should be targeted at this level if not lower. Targeting messages to these and other general audience characteristics is a best practice or risk communication because the audience is more likely to receive and understand the message.

Acknowledge Diverse Levels of Risk Tolerance

One of the consistent themes in efforts to understand risk communication is that various concepts, definitions, experiences, tolerances, and perceptions of risk exist (Slovak et al., 1982). Covello and Johnson (1987) argue that risks are "exaggerated or minimized according to the social, cultural, and moral acceptability" of various activities (p. viii). In addition, people have widely varying capacities to process risk messages, including scientific and technical understandings of risk. Moreover, risk experts often view the pubic as obtuse, uninformed, ignorant, and hysterical about issues of risk and are dismissive of the public's uninformed concerns. When the public senses these attitudes, the complexity of the risk communication problem is compounded (Morgan et al., 2001). Social amplification of risk also can compound the problem by creating even more variance in perceptions of risk. Such amplification, for example, may occur through media reports or through casual social conversations (Kasperson et al., 1988). In general, research has consistently concluded that empirical assessments of risk do not closely correlate with the public's perception of risks.

For many years, social scientists have argued that the perception of reality is inherently constructed through social process (Berger & Luckman, 1966). Social products associated with risk, including understandings, meanings, conventions, and institutions, are produced and reproduced through social processes, including communication processes. As information is created and disseminated about risks, their meaning changes: perceptions of risk may be amplified by social interaction, cultural practices, or the mass media. Age and gender have been shown to impact perceptions of risk. Personal experiences and memories may change the meaning of a particular risk factor.

A best practice in risk communication, then, acknowledges that risk does not merely involve empirical questions, but also social constructions and meaning. Moreover, these social constructions of risk are not necessarily less real or less valid than empirical constrictions of risk. Discounting or denying the validity of personal and cultural experiences of risk enhances the probability that positions will be entrenched and reduces the chances of convergence.

Involve the Public in Dialogue about Risk

One of the emerging perspectives on risk communication emphasizes open and honest public communication about risk so that members of the public can make informed choices. This "risk sharing through risk communication" approach has been driven in part by community right-to-know initiatives. Under the provisions of the 1986 federal Emergency Planning and Community Right-to-Know Act, companies that use or store some 400 chemicals identified as dangerous must report this information to the public. This sharing of information with the public has created an obligation to clarify both the nature and the magnitude of the risk (Kasperson, 1991). Community right-to-know compels a risk sharing approach.

Risk sharing has been a staple perspective in economics for some time. In the insurance industry, for example, risks are often distributed over large groups of subscribers. This spreading of risks over a larger risk pool reduces the level of risk any individual faces. Essentially, a community shares in the risk. In other contexts, such as those involving the environmental protection, risks associated with the location of facilities, or some specific activity-based risks, effective communication can help the general public understand and accept some of the decisional responsibility for the risk. Effective risk communication, then, facilitates decision-making, risk sharing, and further dialogues about the risk (Reynolds & Seeger, 2005).

Leiss (2001) argues that "the public ought always to be involved (through good risk dialogues) in discussions about the nature of risks" (p. xii). These dialogues generally involve collaboration between government, industry and citizens that are open, inclusive, and deliberative. They may take the form of community meetings, working groups, focus groups, or community forums designed to involve diverse stakeholders in discussions about risks. Essentially, they are grounded in multi-directional and ongoing exchanges of information between risk stakeholders.

Risk dialogues contrast with risk monologues, where an organization or agency is the sole voice in providing information and generally makes overly-certain claims about the level and nature of the risk. Leiss (2001) cites many examples of risk communication monologues, including Monsanto's efforts to convince the public that genetically-modified (GM) foods are entirely safe. Through aggressive public relations, marketing, and lobbying, Monsanto overcame all initial objections to GM food. Monsanto also argued that because GM food was completely safe, GM food products need not even be labeled. The objections from consumers being denied even basic information about which foods were GM, and sensitivity of other countries and cultures regarding GM food were initially ignored. The failure of Monsanto to hear and respond to public concerns, and efforts to limit the information to which the public had access precipitated a serious backlash against GM food. Public acceptance of this new food technology has significantly eroded in part because of the perception that concerns were not heard and that these products were being forced on the consumers who had no voice in the decision.

Risk sharing models require open and transparent participatory processes that lead toward shared decision making among government, industry, producers, and the public about risks. In an open process, all risk stakeholders have the opportunity to express their positions. In this context, the opportunity for convergence can emerge. Moreover, there are opportunities for risk communicators to receive feedback and modify their messages to ensure that the communication is effective.

Present Risk Messages with Honesty

Intrinsic to many of the best practices of risk communication discussed so far is the need for information about risks to be communicated in an accurate, honest and even frank manner. Honesty and accuracy are fundamental values of human communication (Seeger, 1996). They are necessary for receivers to make informed, personal choices about risks. The risk sharing model described earlier is predicated on the free flow of information between various stakeholders. In terms of risk, these personal choices are usually framed as issues of self-efficacy, or the ability to take some personal action to reduce a risk factor. Self-efficacy is generally a very effective strategy in risk communication in that it reduces the perception of powerlessness and promotes personal responsibility and action in managing risk. Gordon (2003) found that effective risk messages about foodborne illness occurred when perceptions of risk were elevated and some guidance was provided about what personal actions can be taken to reduce the risk. Given the fact that many issues of food safety involve food-handling on the consumer level, self-efficacy is particularly important.

Significant impediments exist to open, honest, and frank communication. Many risk communication messages, for example, are essentially manipulative, and designed to generate compliance or agreement with a particular view of what is risky (Morgan et al., 2001). Many risk communication campaigns are designed to overwhelm any objections by drowning out critics or by portraying them as irrational and uninformed. In some cases, managers are fearful that open and honest discussions of risk will promote, or further, irrational concerns and fears and enhance legal liability. Sometimes, managers believe that open and honest risk communication may involve disclosing trade secrets. In still other instances, there simply may be insufficient information about a risk and acknowledging this uncertainty may create heightened public concern.

As noted earlier in the case of Monsanto's efforts to generate public acceptance of GM foods, ignoring an issue of risk, discounting public concerns, or overwhelming objections with a barrage of strategic messages does not remove the issue from the public domain. Models of issue management indicate that, in these cases, issues will simply arise at some later date when another group or agency raises the issue. While open and honest risk communication may appear to be ineffective in the short term; in the long term, these approaches foster and prolong accommodating relationships with a variety of stakeholders.

Meet Risk Perception Needs by Remaining Open and Accessible to the Public

Closely associated with honest communication is accessibility and openness. Heath et al. (1998) note that openness is necessary for two-way, symmetrical communication between organizations and their stakeholders. Accessibility and openness enhance the public's perception that they are fully informed about a risk and that they are partners in sharing the risk. In addition, the public may have specific needs for particular forms of risk information. Accessibility has many dimensions, including the form of information, receiver characteristics, location, and channels of communication.

Risk information most often takes a highly technical form, characterized by scientific jargon and technical terminology. In most cases, information in this form is simply not accessible to lay persons or the general public. While subject matter experts often believe that technical information is the most accurate, in many cases it simply frustrates members of the public who are trying to understand the risk. Many members of the public may come to feel they are being talked down to and patronized. Similarly, receiver characteristics (reading and education level, abilities, culture, age) may influence the ability to access information. Research indicates that individuals from some cultures prefer that information come from trusted cultural agents as opposed to established authority figures or technical experts. Language barriers often create profound challenges to accessibility and openness. Location issues are significant barriers to accessibility for many communities. Finally, the medium of communication may be more or less familiar and accessible to various members of the public. Information may be posted in places that are more or less accessible. Despite the ubiquitous nature of communication technology, not everyone uses the Internet. Accessibility, then, is highly variable by audience, information, and channel. Simply making information available on a web site or in the fine print on a product label does not ensure accessibility and openness.

In addition, openness and accessibility denotes a general attitude on the part of the organization or agency about issues of risk. A closed or inaccessible stance can create the impression that there is something to hide. In these cases, members of the public and the media often become more aggressive in their efforts to access information. In contrast, openness can help build trust and promote a sense of collaboration and risk partnership. As a best practice, then, meeting the informational needs of the public and remaining open and accessible serves a variety of goals including promoting self efficacy, building trust, and ultimately facilitating convergence.

Collaborate and Coordinate About Risk with Credible Information Sources

A final best practice concerns how risk communicators interact with each other. When considering any particular risk, a variety of organizations, agencies, and groups will likely serve as sources of information. In the case of a food related risk, for example, food producers, industry groups, several federal and state agencies, and consumer groups can be expected to serve as sources of information. Risk communication, in these cases, can be significantly compromised if these agencies and groups offer contradictory and inconsistent messages. Consistent messages can help develop a more coherent and effective public understanding of risks. Alternatively, inconsistent messages can sow confusion and create competing interpretations of risk.

Drabek and McEntire (2002) define coordination as, "collaborative processes through which multiple organizations interact to achieve common objectives" (p. 199). Collaboration requires that organizations reach across their boundaries, which are sometimes rigid and often lack permeability. Coordination may take many forms, including bureaucratic, cultural, and informational. Bureaucratic coordination is rule and procedure-based, and is often imposed by agency regulations. Cultural coordination occurs when organizations have similar operations and values, such that trust develops more quickly among them. Finally, informational coordination occurs when organizations regularly exchange information, allowing one to know what the other is doing. Modern information systems, such as radio frequency identification devices (RFID), have facilitated this form of coordination.

Coordination of risk assessments and risk communication strategies requires information sharing and establishing networks of working relationships between groups and agencies. Establishing these relationships necessitates overcoming institutional, cultural, and political boundaries. Significant barriers exist between regulatory agencies and industry groups. Traditionally, these groups have mistrusted one another, and cooperation and collaboration, including sharing information, correspondingly has been limited. The adoption of radio frequency identification technology for tracking livestock, for example, has been met with significant resistance due in part to mistrust between regulatory agencies and producers (Veil, 2006). In the food industry, the need for coordination has been enhanced by industry integration and globalization of both markets and production. In the case of GM foods discussed earlier, disagreements between U.S., European Union, and Canadian regulatory agencies fueled the debate over the safety of GM crops.

Overcoming institutional and cultural barriers, and mistrust is necessary to create consistency in risk messages. Open communication and information sharing can help clarify where risk perceptions diverge and identify points of convergence. The outcome may not be universal agreement about risks, but convergence around the general parameters of risk.

Summary

These best practice strategies of risk communication are not designed to function as distinct steps or isolated approaches. Rather than being mutually exclusive, they serve to complement one another and create a coherent approach to confronting risk communication problems. These best practices are grounded in a core set of communication values and promote the larger goal of creating convergence and understanding around issues of risk.

The nine best practices of risk communication are in many ways idealized positions that may not be available in all risk communication circumstances. Short term economic pressures can create significant impediments to open and honest risk sharing approaches. The drive to innovate and become more competitive forces the adoption of new technologies, methods, and products before the risks can be fully explored and communicated. Institutional and cultural barriers can be rigid impediments to the exchange of information. Many organizations find that these approaches need to be implemented incrementally, so that channels of communication and trust may develop over time between stakeholders. An incremental approach creates the opportunity for feedback and the adjustment of communication strategies to improve effectiveness.

Risk communication is a highly complex and dynamic process which creates profound obligations and challenges for organizations, agencies, industry, and consumer groups. Among the most important dynamics of risk communication is that failure to communicate risk effectively can ultimately increase the risks faced by organizations, agencies, and the publics. In the end, seeking convergence about risk through the nine practices described here is itself an important risk management strategy.

1. Parts of this chapter are based on: Seeger, M. (2006). Best practices in crisis communication: An expert panel process. *Journal of Applied Communication Research, 34,* 232–234.

References

Ahmed, P. K., & Rafiq, M. (1998). Integrated benchmarking: A holistic examination of select techniques for benchmarking analysis. *Benchmarking for Quality Management and Technology, 5*(3), 225–242.

Berlo, D. K. (1977). Communication as process: Review and commentary. In B. D. Ruben (Ed.), *Communication Yearbook 1* (pp. 11–27). New Brunswick, N.J.: Transaction.

Berger, P. L., & Luckman, T. H. (1966). *The social construction of reality: A treatise in the sociology of knowledge*. New York: Anchor Books.

Cohen, M. D., & Sproull, L. S. (1996). *Organizational learning*. Thousand Oaks, CA: Sage.

Conrow, E. H. (2003). *Effective risk management*. Reston, VA: American Institute of Aeronautics & Astronautics.

Covello, V. T., & Johnson, B. B. (1987). *The social and cultural construction of risk*. Boston, MA: Dordrecht.

Dozier, D. M., & Broom, G. M. (1995). Evolution of the managerial role in public relations practice. *Journal of Public Relations Research, 7*(1), 3–36.

Drabek, T. E., & McEntire, D. A. (2002). Emergent phenomena and multiorganizational coordination in disasters: Lessons from the research literature. *International Journal of Mass Emergencies and Disasters, 20*, 197–224.

Gilchrist, J. A., & Browning, L. D. (1981). A grounded theory model for developing communication instruction. *Communication Education, 30*(3), 273–277.

Glaser, B. G., & Strauss, A. (1967). *The discovery of grounded theory: Strategies for qualitative research*. Chicago, IL: Aldine.

Gordon, J. (2003). Risk communication and foodborne illness: Message sponsorship and attempts to stimulate perceptions of risk. *Risk Analysis, 23*(6), 1287–1296.

Heath, R. L., Seshadri, S., & Lee, J. (1998). Risk communication: A two-community analysis of proximity, dread, trust, involvement, uncertainty, openness/accessibility, and knowledge on support/opposition toward chemical companies. *Journal of Public Relations Research, 10*, 35–56.

Kasperson, R. E. (1991). *Communicating risks to the public*. Norwell, MA: Kluwer.

Kasperson, R. E., Renn, O., Slovic, P., Brown, H. S., Emel, J., Goble, R., Kasperson, J. X., & Ratick, S. (1988). The social amplification of risk: A conceptual framework. *Risk Analysis, 8*(2), 177–187.

Kyro, P. (2004). Benchmarking as an action research process. *Benchmarking, 11*(1), 52–60.

Leiss, W. (2001). *In the chamber of risks: Understanding risk communication*. Toronto: McGill-Queen's Press.

Littlefield, R., Cowden, K., Farah, F. McDonald, L. R., & Sellnow, T. (2006). *Ten tips for risk and crisis communicators when working and conducting research with Native and New Americans*. Fargo, ND: Institute for Regional Studies.

Morgan, M. G., Fischoff, B., Bostrom, A., & Atman, C. (2001). *Risk communication: The mental models approach*. New York: Cambridge University Press.

Nicotera, A. M. (1983). Beyond two dimensions: A grounded theory model of conflict-handling behavior. *Management Communication Quarterly, 6*(3), 282–306.

Reynolds, B., & Seeger, M. W. (2005). Crisis and emergency risk communication as an integrative model. *Journal of Health Communication Research, 10*(1), 43–55.
Seeger, M. W. (1996). *Ethics and organizational communication*. Cresskill, NJ: Hampton Press.
Seeger, M. W. (2006). Best practices in crisis communication: An expert panel process. *Journal of Applied Communication Research, 34*(3), 232–234.
Slovic, P., Fischoff, B., & Lichtenstein, S. (1982). Facts versus fears: Understanding perceived risk. In D. Kahneman, P. Slovic, & A. Tversky (Eds.), *Judgment underuncertainty: Heuristics and biases* (pp. 463–489). Cambridge: Cambridge University Press.
Veil, S. R. (2006). *Crisis communication and agrosecurity: Organizational learning in a high-risk environment*. Unpublished doctoral dissertation, North Dakota State University.

Chapter 3
Multiple Audiences for Risk Messages

A house full of people is a house full of different points of view.
(Maori Proverb)

As we have demonstrated in previous chapters, risk is pervasive today due to ever-increasing levels of uncertainty about all aspects of life. Thus, risk communicators are continually striving to gain the public's adherence to specific strategies with the potential to ward off crises. The public, or audience, often is labeled as any group of individuals who hear and respond to messages directed toward them. The term, *general public*, reflects the perception that the whole group is included. However, audiences are far more complex than the term *general* implies. Within any given audience, elements representing diverse perspectives exist that may or may not be receptive to the specific strategies presented. Often, when the common single spokesperson model is used for communicating during a risk or crisis situation, the general public receives the focus of attention and diverse perspectives are not acknowledged. This can exacerbate the risk or crisis situation.

This chapter identifies the need for including multiple audiences or publics in the development and transmission of risk messages. We begin with a discussion of the current emphasis on sender-oriented risk communication and the need for an audience-centered approach. Then the concept of multiple audiences is provided with practical applications for stakeholders, as well as those who craft and receive risk messages. Spheres of ethnocentricity (Littlefield & Cowden, 2006) are introduced as a way to understand the varied responses of multiple publics to risk messages. We next examine some dimensions of culture affecting the receptivity of particular audiences to risk and crisis messages through the single spokesperson model and a variation where scripted messages are relayed through cultural agents. Finally, applications of the principle of interacting arguments through the lens of the culturally-sensitive approach, as well as some benefits of following the culture-centered approach to risk communication are presented.

T. L. Sellnow et al. *Effective Risk Communication*

DOI: 10.1007/978-0-387-79727-4_3

Sender Versus Audience-Centered Communication

The conceptual framework for this chapter focuses on risk and crisis communication research and the need to include a broader understanding of multiple audiences or publics who are receiving risk and crisis messages. The emphasis in most of the risk and crisis communication research focuses on the viewpoint of *elites* in organizations and groups as they move through the pre-crisis, crisis, and post-crisis stages (e.g., Barton, 1993; Benoit, 1995, 1997; Covello, 1992, 2003; Heath, 1997, 2001; Lindell & Perry, 2004; Seeger et al., 1998, 2003; Slovic, 1986; Weick, 1988, 1995). Elites represent those who are in positions of leadership to plan and manage the communication and actions during the pre-crisis, crisis, and post-crisis stages.

In contrast to the robust research focusing on the senders of risk and crisis messages, few studies have explored audience perceptions of these messages. Most of what is known about message-testing comes from studies of advertising and public relations, emphasizing the importance of involving people in order to discover more effective ways to change attitudes and behaviors (e.g., Fink, 1986; Leanna et al., 1992; Leitch & Neilson, 2001; McMahan et al., 1998). While scholars suggest that audience analysis be conducted when constructing risk and crisis messages, many regard audience characteristics (ethnicity, economic status, education, family size and status, household structure, information retrieval, language mastery, neighborhood, and technology) as variables affecting the outcomes sought by elites within the organizations and groups preparing the risk and crisis messages, as opposed to factors shaping the construction of the risk or crisis message for multiple publics (Coombs, 1998, 1999, 2007; Lindell & Perry, 2004; Rowan, 1991; Slovic, 1986; Tierney, 1999).

Inclusion of culture as a variable is not new to risk and crisis communication research. However, most scholars have defined culture from the sender-oriented perspective. That is, cultural groups are viewed as *we-they* or *us-them*. Chess (2001) identified culture as a variable that should be taken into account as an organization perceives risk and crisis. Seeger, Sellnow and Ulmer (2003) extend this, suggesting an application to changing cultural dynamics: "As an organization monitors the external environment, it must continually reformulate risk-related messages" (p. 204). Lindell & Perry (2004) offer an alternative perspective, defining culture as, "shared beliefs that encompass both people's interpretations of the world and their notions of how one deals with (responds to and attempts to influence) that world" (p. 17). Despite this more audience-centered definition, past research in risk and crisis communication provides little insight into the process of *how* multicultural publics perceive risk and crisis messages.

Identification of Multiple Audiences

Within any given audience, multiple groups may exist who perceive messages differently. The ConAgra example in Chapter 9 extrapolates this reasoning. This makes the identification of these elements not only useful, but essential if risk communicators are to gain audience members' compliance to specific strategies for preventing or managing crises. To provide a framework for identifying these audiences, a theoretical perspective is offered that clarifies the difference between general audiences and the diverse elements included within them.

The Universal and Particular Audiences

An understanding of audience types is useful when considering how diverse publics perceive risk messages. Perelman and Olbrechts-Tyteca (1958/1971) provided a theoretical framework for differentiating between audience types and framing arguments to gain compliance. Specifically, their discussion of the *universal* and *particular* audiences illustrated factors complicating effective risk communication with multiple publics.

Perelman and Olbrechts-Tyteca (1958/1971) define these audiences as follows: "The [*universal* audience] consists of the whole of mankind, or at least, of all normal, adult persons. . . . The second consists of the single *interlocutor* whom a speaker addresses in a dialogue" (p. 30). The universal audience represents those who are like the speaker: reasonable and rational individuals. The speaker believes that "if the message makes sense to me, it should make sense to others." While this construction is ethnocentric because the speaker's view of the audience dominates, it reflects the outer limits of what should be known or acted upon in a given situation: in this case, a risk situation.

In contrast to the universal audience, *particular* audiences may be constructed variously. Initially, the speaker may consider the particular audience to be, in fact, the universal audience. Perelman and Olbrechts-Tyteca (1958/1971) conclude, "It none the less remains true that, for each speaker at each moment, there exists an audience transcending all others, which cannot easily be forced within the bounds of a particular audience" (p. 30). Slogans or platitudes such as, "Wash, Separate, Cook, Chill," designed to warn people about how to handle uncooked meat, represent messages where the bounds between universal and particular dissolve. However, as the risk communicator shapes messages to appeal to diverse elements, target or particular audiences soon emerge from the universal.

Identification of Stakeholders

The audiences for risk and crisis messages are comprised of stakeholders, defined in Chapter 1 as "any person or group of persons whose lives could be impacted by a given risk." The term stakeholders reflects the point of view of organizational or community elites who consider elite individuals and groups as receivers of their risk and/or crisis messages. In this context, elites expect that their messages to other elite stakeholders and key publics will create awareness, explain the risk or crisis, gain agreement regarding a specified course of action, and motivate the message receivers to do something (Rowan, 1991).

However, acknowledgment of stakeholders and key publics also suggests a position of privilege for these individuals and groups at the expense of others who are not stakeholders or regarded as key publics. For example, prior to Hurricane Katrina, crisis managers in New Orleans spent considerable time discussing emergency evacuation routes (see Chapter 6). Following the devastation, however, these same managers acknowledged they had not given much thought to the fact that most of the poorest residents relied upon public transportation. Their focus had been on those with the economic resources to have their own cars. Quite often, those not considered as key publics represent different ethnic or cultural groups, socio-economic levels, or educational backgrounds.

Message Construction

In the context of risk, despite the National Research Council's characterization of risk communication as interactive, spokespeople typically script their messages for the general public or, in their minds, the universal audience. These risk communicators consider that if certain facts or truths are presented, the automatic compliance of those hearing the message will be gained. Because they believe that stakeholders will accept their message, a single spokesperson is used to provide this perspective, reflecting the model of dominance introduced in Chapter 1.

As single interlocutors, the risk communicator considers individual hearers to be "an incarnation of the universal audience" (Perelman & Olbrechts-Tyteca, 1958/1971, p. 37). However, just as there are many individual hearers within the universal audience, there are also multiple groups or publics. Thus, each individual or particular audience may respond differently to the single risk message. For instance, when scientists or elites communicate with others of like mind or experience, they may expect agreement with or acceptance of their position. But there will be some among those excluded from this elite group who are uncertain or indifferent to the messages.

Risk communicators must construct their messages to take into account the audience's composition. In the context of interacting arguments, this construction helps the risk communicator to be more persuasive as the multiplicity of arguments resonates differently with the various publics receiving the messages. Similarly, multiple publics within the particular audience require preparing messages in accordance with how these publics are conditioned to receive them. Perelman and Olbrechts-Tyteca (1958/1971) explain: "Knowledge of an audience is also knowledge of how to bring about its conditioning, as well as of the amount of conditioning achieved at any given moment of the discourse" (p. 23). In the context of risk messaging, conditioning agents might include visual or auditory signifiers to remind or catch the publics' attention. For example, using an official logo to signal authority, the American flag to suggest loyalty, recognized warning symbols to focus attention, patriotic music to stimulate emotion, or warning sirens to stimulate immediate action may provide the conditioning that is a element of the interacting arguments identifying the textual message and the immediacy of the conditioning component.

Creators of risk messages often represent elite groups who perceive themselves as convincing to all rational beings. In essence, "the elite audience sets the norm for everyone" (Perelman & Olbrechts-Tyteca, 1958/1971, p. 34). This leads the risk communicator to assume that hearers will accept the premise of risk upon which their message is constructed. However, cultural variables may counteract this assumption from the outset, rendering the message ineffective in accomplishing its intended result. To illustrate, when some Native American groups are warned of potential risk they interpret the situation as one where the actual risk and crisis already exists. They believe that "if the problem or risk/crisis situation didn't already exist, someone wouldn't be telling us about it now." This belief is based upon Native Americans' historic relations with the U.S. Federal Government, when potential negative situations became reality. As a result, with some Native Americans, risk communicators would be better off waiting until the crisis is at hand before providing specific strategies for coping with the situation.

The risk communicator's goal is to gain adherence from those who hear the message: "An efficacious argument is one which succeeds in increasing this intensity of adherence among those who hear it in such a way as to set in motion the intended action . . . or at least in creating in the hearers a willingness to act which will appear at the right moment" (Perelman & Olbrechts-Tyteca, 1958/1971, p. 45). This necessitates risk communicators using content and style consistent with that goal. However, the nature of the content and style preferred by the universal and particular audiences may not match. When that happens and multicultural audiences do not react or comply with the elite's message, the audiences are blamed and labeled by elites as "stupid" or "abnormal" (p. 33). "It's their own fault that bad things happened to them," the communicators suggest. "If they were smart, they would do as I say."

The risk communicator's focus is on the universal audience, with the goal of convincing hearers that a particular response is required. In the context of the universal audience, vague and abstract messages are a prerequisite. However, due to the cultural variables shaping a particular audiences' perspectives, providing specific instructions predicated on the elite's perspective of what all rational people would accept as fact is not effective. Entente is impossible when specific instructions are provided in the universal context. Rather, a shift to the approach enabling each particular audience to participate in the process of establishing interacting arguments is more productive for the risk communicator.

Risk communicators who do not take into account particular audiences and interacting arguments cannot use overarching universal appeals because publics require responses appropriate to the temporal and local circumstances. Providing a universal set of instructions in preparation for a potential crisis situation is ineffective for audiences that may require multiple instructions for action. Value hierarchies assumed to be universal, such as the right of humans to control nature, fail because they can be refuted by alternate value systems, such as an existential or fatalistic model suggesting the future has already been determined. The goal of gaining adherence by all to the recommended course of action, by appealing to the universal audience, may not be realized because what one hearer accepts, another rejects.

Risk communicators who engage multiple communicators working through entente with their particular groups can be effective. Risk communication in this context is more productive because representatives of particular audiences address their own groups. The specific actions recommended by a particular audience for its members gains adherence because the suggested actions are derived from within the culture itself. The particular value assigned to the proposed action is accepted because it resonates with the cultural group. The goal of multiple risk communicators seeking adherence by their particular audiences to the recommended course of action is realized because the particular has influenced the form the course of action has taken.

Spheres of Ethnocentricity

Despite risk communicators' efforts to gain adherence to their warnings, multiple audience members' attention to risk messages is predicated on the *spheres of ethnocentricity* (Littlefield & Cowden, 2006) that influence their perceptions of what is or is not pertinent to their circumstances and worthy of their attention. Spheres of ethnocentricity may be best described as concentric circles of family, community, region, state, nation, and world going out from an individual as the center of the sphere (Fig. 3.1).

The closer to the center of the spheres, the more focus the individual has on the risk or crisis messages being sent and the necessary strategies available to respond to the crisis. The farther out from the center, the less interest the individual may have in the message and the less direct control the communicator has in gaining adherence. These spheres of ethnocentricity affect how much attention the individual or group will pay to the risk or crisis message. If the risk or crisis affects an individual or his/her family, the action taken will reflect the intensity of feeling at that level. If a risk or crisis exists in the world, the outermost concentric circle, less intensity may be reflected in the individual's actions.

How individuals interpret, understand, or respond to risk messages is based upon their place within the spheres. Adherence by particular audiences to specific response strategies is based upon the dominant sphere in which the individuals within that audience find themselves. Thus, to establish entente, risk communicators must understand the reality of particular audiences. This requires convergence between the risk communicator and the intended audience, or systemic convergence if multiple audiences are involved in the risk or crisis situation.

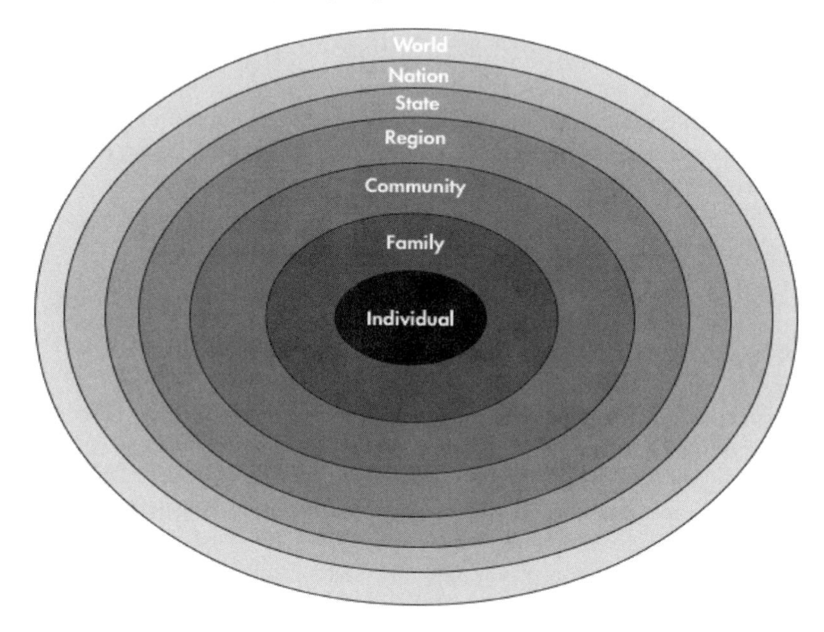

Fig. 3.1 Spheres of ethnocentricity

Dimensions of Culture Affecting Receptivity

The concept of culture interests risk communicators for many reasons. Cultural groups may view differently what constitutes a risk or crisis situation; they also may respond differently to risk messages and crisis situations. Hurricane Katrina provides a vivid example of a situation where cultural variables in part may have been responsible for the failure of thousands of African Americans and lower-income residents to respond as emergency managers anticipated to the risk and crisis messages. The reasons vary, but two factors may have contributed to the mindset of those who sheltered in place: the crisis's timing (on payday at the end of the month) and lower income residents' reliance on mass transit for transportation.

Culture needs to be considered in risk communication theory and practice. At present, culture is regarded as a collection of static variables pertaining to particular groups' values: how they think, and what they do. The emphasis in the *status quo* for risk communicators is on configuring the already-existing message and process for delivering that message to fit cultural characteristics and preferences of target groups. Expertise remains in the hands of dominant external risk and crisis experts who determine the objectives and relevant cultural characteristics, configure the message to fit the characteristics, and evaluate their effectiveness based upon their own criteria.

The Impact of Culture on Communication

The assumption behind the traditional risk and crisis communication model is that one person presenting one message will be more effective in supplying information to the general public about how to respond to a crisis situation (Fig. 3.2).

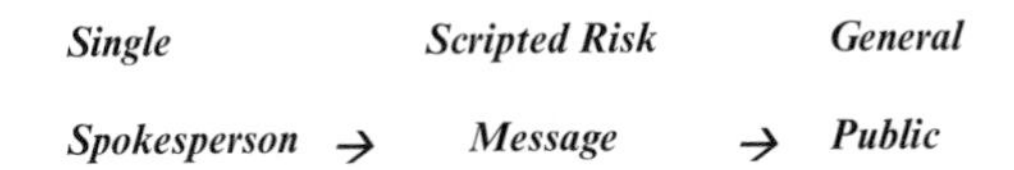

Fig. 3.2 Single spokesperson model of risk communication

This speaker-centered approach draws strength from what Klopf (1991) terms *projective cognitive similarity*, or the belief that "the person with whom we are talking perceives, judges, thinks, and reasons the same way we do" (p. 223). This perspective provides some assurance that the public will receive a consistent message in times of risk and crisis, much like what Rogers (2003) suggests was necessary for the successful dissemination of any new information into a social system. While this theoretical position appears logical, in reality it is not practical because there is not just one public receiving a crisis message. Instead, there are multiple publics represented by a wide range of ethnic and cultural groups who are asked to receive the cross-cultural message uniformly and respond accordingly. Unfortunately, due to socio-cultural variables, their responses are often far from uniform.

While a crisis situation may benefit from a centralized information source, an alternative way to reach multiple publics may involve multiple spokespersons drawn from the affected particular audiences or cultural groups (Fig. 3.3).

	Cultural Agent 1→	***scripted message→ specific public***
Single	***Cultural Agent 2→***	***scripted message→ specific public***
Spokesperson →	***Cultural Agent 3→***	***scripted message→ specific public***
Provides	***Cultural Agent 4→***	***scripted message→ specific public***
	Cultural Agent …→	***scripted message→ specific public***

Fig. 3.3 Multiple spokesperson model of crisis communication

Scholars in intercultural communication recognize the diversity of the multiple publics and have identified a number of factors that affect how culturally diverse groups send and receive messages. These factors range from macro to micro in scope, but all can change the way a crisis message is received. While the number of cultural factors affecting communication is staggering, we use Sarbaugh's (1979) four general categories of a taxonomy in discussing culture's effects on communication because they offer clarity when applied to crisis messages communicated to diverse publics. The four categories are code systems, perceptions about relationships and intent, knowing and accepting normative beliefs and values, and world view.

Code Systems

Code systems, or language variations, present a major challenge for effective communication involving multiple audiences. Distinctive features of language, different language rules, and alternative functions of languages are noted as cultural variables affecting communication. Ting-Toomey and Chung (2005) suggest that language is one culture-specific variable "governed by the *multilayered rules* developed by members of a particular sociocultural community" (p. 141). That being the case, the "arbitrariness, abstractness, meaning-centeredness, and creativity" of language makes communication across cultures a complex undertaking (p. 141).

Drawing on the work of Ting-Toomey (1989), Fong (2006) identifies three approaches to understanding how culture influences language and communication. The developmental approach focuses on how language affects the way people think; the interactional approach identifies how communication styles and norms in different cultures affect the way people interact; and the social-psychological approach investigates language choice in multilingual communication contexts, particularly first-language or second-language usage in majority and minority groups within communities (pp. 216–217).

Speakers' verbal styles also influences how messages are perceived. Some scholars have noted the specific language characteristics of various cultural groups. For example, Gudykunst and Ting-Toomey (1988) identify four variations of verbal communication styles affecting levels of understanding. The direct versus indirect style involves "the extent speakers reveal their intentions through explicit verbal communication" (p. 100). For some cultures, an elaborate style affects the level of "rich, expressive language" used to communicate a message (p. 105). Another style involves the personal rather than the contextual. Verbal style uses "certain linguistic devices to enhance the sense of 'I' identity, while verbal contextual style uses certain linguistic symbols to emphasize the sense of 'role' identity" (p. 109). Finally, the instrumental style in terms of the language used, is characterized as "sender-oriented" rather than "receiver-oriented" (p. 112).

Other scholars have provided insight about code systems and language choices more specific to particular cultures. Klopf (1991) offers details about the language patterns of international cultural groups such as Japanese, Mexican, Arabs, Chinese, and Germans. Within the United States, Neuliep (2003) identifies traits of Spanglish (the language of Hispanic Americans), Black English and Ebonics, and the languages of the Amish and the Hmong.

These various perspectives on differing language styles and code systems offer specific advice for spokespersons in times of risk and crisis. Since individual cultures have specific elements associated with language, the use of one crisis message transmitted across cultures is ineffective in establishing entente and motivating individuals to respond appropriately to the situation. Unless these variables are taken into account, language differences and styles of communication are likely to increase misunderstanding or result in non-compliance to crisis messages. With only a single spokesperson presenting one message cross-culturally, individuals in the different cultural groups may to respond as directed.

Perceived Relationship and Intent

The way individuals view their relationship with members of other cultures and the corresponding communicators' demonstrated intent can affect how messages are received. Hofstede (1991) identifies four broad cultural patterns that influence how individuals perceive each other and respond to intercultural communication. These include power distance, individualism and collectivism, uncertainty avoidance, and masculinity and femininity.

With power distance, individuals perceive messages differently according to their place within the social hierarchy, their relationship with authority figures, and the intent of the person sending the message. The degree to which individuals act independently, rather than as members of a group, can affect their receptivity to messages aimed at encouraging them to follow directions for the common good rather than for their own benefit. In a culture with higher levels of tolerance for ambiguity and uncertainty, a risk message's call for action may have less urgency. Qualities associated with masculinity or femininity in cultures may also affect people's reception of a risk message that appears more assertive than nurturing in intent. The effect of variables is evident as single spokespersons establish relationships based upon authority and power, distancing them from the different communities, and providing no guarantee that the publics will respond positively by following the official directives.

Knowing and Accepting Normative Beliefs and Values

Another aspect of intercultural communication involves the communicators' receptivity to knowing and accepting each other's normative beliefs and values. When the beliefs and values are known and accepted, positive intercultural communication is the result. If beliefs and values are neither known nor accepted, misunderstanding and distrust will occur. As a dimension of knowing and understanding these differences, Hall (1976) proposes that communication is divided into high context and low context systems.

Depending upon an individual's cultural orientation, the manner of communication may be directly influenced. For example, in a high context culture, beliefs and values are understood and accepted without explicit explanation. In a low context culture, explicit information about the beliefs and values must be shared if there is to be knowledge and acceptance. Ting-Toomey and Chung (2005) classify these high and low context communication patterns as follows. High context patterns reflect collectivist values (all understand), spiral logic (all thought is connected), indirect verbal style (no need to speak the obvious), understated or animated tone (nonverbal communication dominates), formal verbal style (demonstration of respect), and verbal reticence or silence (unwillingness to confront). Low context patterns include individualistic values (self-focused), linear logic (one step follows another), direct verbal style (willing to ask and tell), matter-of-fact tone (common expectation to get more information), informal verbal style (no one commands more respect than another), and verbal assertiveness or talkativeness (behavior demonstrates demand for information) (p. 170).

Harris and Moran (1991) suggest a culture's characteristics influence normative beliefs and values. Characteristics specifically related to risk or crisis communication include sense of self and space, food and feeding habits, time and time consciousness, values and norms, beliefs and attitudes, mental process and learning, and work habits (pp. 206–211). Each of these characteristics has the potential to affect how receivers of risk or crisis messages know and accept them. For example, if a culture is informal and flexible, members may not respond to formal messages providing explicit instructions about how to behave when dealing with a risk or crisis. Beliefs and values pertaining to food safety vary, since every culture differs in the way food is selected, prepared, presented, and eaten. The way individuals respond to time differs by culture; an immediate crisis in one culture may not carry the same urgency in another. Cultural values differ and norms of behavior regarding what is and is not considered acceptable vary. Related to these values are beliefs that shape how a risk or crisis is viewed. Specifically, if a group believes that a supernatural power controls their destiny, taking steps to protect the group from a food-related crisis may be futile. The way individuals organize and process information is also reflected in their ability to respond to real or potential threats to their safety. Finally, the practical importance of work to sustain daily life may make it difficult for individuals to change their behavior when confronted with a crisis.

In brief, for the risk or crisis communicator, the more that is known about the receiving publics' normative beliefs and values, the greater the chances for those beliefs and values to be reflected in the message. As every characteristic of a culture has the potential to influence and reflect beliefs and values, the need for attention to this area of cultural variability is clear.

Worldview

Communication between cultures is further complicated by the various ways people perceive and act in the world around them. Perceptions about the nature of life, the purpose of life, and the human relationship to the cosmos contribute to an individual's world view. Sarbaugh (1979) suggests that the nature of life refers to how humans experience their reality: "Questions of mind, body, and soul are aspects of the beliefs about the nature of life" (p. 43). The purpose of life involves how people should direct their efforts as they experience their lives. There may be as many different purposes identified as there are individuals to name them, ranging from striving to control as much material wealth as possible to doing everything possible to manage risk.

The relationship of humans to the cosmos has something to do with how our relationship with nature and the spiritual world is viewed. For example, humans may be subjugated to nature, be equal with nature, or attempt to dominate nature. From another perspective, individuals may feel controlled by the crisis, feel up to the challenge of the crisis, or want to control or manage the crisis. Some cultural groups may view the crisis situation as beyond control while another group may consider itself master of its destiny.

Another dimension of world view takes into consideration a culture's religious or spiritual aspects. Samovar and Porter (2001) identify six major religions affecting roughly 90% of the world's populations: Buddhism, Christianity, Confucianism, Hinduism, Islam, and Judaism. Each of these has different sacred writings, authority figures, rituals, speculation, and ethics that shape how their followers identify with and understand messages related to their well-being.

Some scholars suggest that value orientations contribute to worldview and have a powerful influence on the way members of a culture perceive and respond to communication (Kluckhohn & Strodtbeck, 1961). Klopf (1991) believes that the way members of a culture perceive, think, and speak is influenced by the way they view the world around them. Condon and Yousef (1975) examine the value orientations of different cultural groups and identified six dominant themes that helped to explain how the hierarchy of values held by particular cultural groups complicated communication with multiple publics. These themes include: how the self is identified, the role of the family, societal expectations, elements of human nature, the relationship of people to nature, and the role of the supernatural. How individuals actually see themselves in relation to the other values within the themes contributes to what Schwartz (1992) refers to as value priorities.

In short, code systems differences, misperceptions of relationships and intent, conflicting points of view regarding normative beliefs and values, and differing worldviews provide ample support for risk communication scholars rethinking how spokespeople communicate with multiple publics. In response to these confounding differences, the intercultural communication literature provides support for the culturally-sensitive model when communicating with different publics. Among the findings are the following claims drawing support from the taxonomy developed by Sarbaugh (1979):

- The more diverse the publics, the less efficient a single spokesperson will be in communicating a risk or crisis message.
- If spokespeople do not share a common code system (both verbal and nonverbal) or have a mechanism for translating into a common code with the particular publics, then the desired goal of communicating the seriousness of a risk or crisis becomes less likely.
- If the relationship between the spokesperson and public is perceived to be friendly and helpful, the participants more likely will respond positively and follow the instructions for dealing with the crisis. Conversely, if the relationship has a hostile, dominant, disruptive tone, the less likely the participants will respond as instructed.
- If the intent shifts from helping to disrupting, there will be resistance to the risk or crisis message.
- The more heterogeneous the relationship between the spokesperson and the diverse public, the greater the probability for communication difficulties.

- If two participants have different patterns of beliefs and behaviors, they will respond differently to communication messages.
- If participants do not know and accept the normative beliefs and behaviors, the difficulty in carrying out the transaction increases and probability of communication breakdown increases.
- The greater the difference in worldview, the more difficult it will be for a single spokesperson to convey the severity of the risk or crisis.

Applications of Interacting Arguments

Cultural considerations may be applied to the principle of interacting arguments in risk communication in a number of ways. In the case of *convergence* between the risk communicator and a particular audience, both share in the process so the script is developed integrating knowledge from both parties, the message is created, and the appropriate channels or manners for delivering the message are identified. Due to the nature of *congruence* and the multiple audiences who may hear a risk message, the single-mindedness expected in risk situations is unlikely. Because particular audiences may not have access to the same kinds of knowledge as risk communicators, the total disregard of one or the other body of knowledge in favor of the other–*mutual exclusivity*–is counterproductive and unlikely to occur. The culturally-sensitive approach may result in *dominance*, since the risk communicator has the power to decide which cultural variables will apply in an effort to gain a particular audience's adherence. Under the assumption that particular cultural variables are being attended to, adherence may occur despite the presence of other variables. What is likely in risk situations is *systemic multiple convergence* when more than one cultural group is involved in developing and transmitting messages to different audiences or publics.

Gudykunst and Ting-Toomey (1988) developed a model for understanding the influence of cultural variability on communication in interpersonal settings. In the model, language, ecology, history, and communication affect the socio-cultural variables influencing social cognitive processes, situational factors, dimensions of communication, and habits of behavior. These elements lead to understanding and intention, ultimately producing communication with another person. The facilitating conditions stemming directly from the situational factors ultimately affect the communication. This model has utility for crisis communication in cross-cultural contexts because to communicate crisis messages effectively to multiple publics with differing cultural backgrounds, greater attention must be placed on the facilitating conditions (or the person's knowledge of those socio-cultural variables) influencing communication.

The culturally sensitive approach (Dutta, 2007) reflects the belief that in order to communicate effectively, communicators must understand the characteristics of the cultural groups with whom they wish to engage and alter existing communication practices to engage the cultural groups: "Research using the culturally-sensitive approach seeks to figure out the most important characteristics of a culture that would lead to the development of successful health messages" (p. 308). In the case of risk and crisis communication, communicators select the cultural variables perceived to be most salient for a particular group and present the risk or crisis messages with the expectation of the cultural groups' compliance because the "underlying cultural dimensions" have been incorporated in the message (p. 308). In Dutta's discussion, culturally-sensitive communicators consider cultural categories to be static and the message's intent to be compliant with particular behaviors. The culturally-sensitive approach relies on the communicator to determine which cultural variables will help to persuade individuals within the cultural group to respond in a way that benefits them and is appropriate to the risk or crisis situation. The risk communicator secures the adherence of the intended audience by constructing a culturally sensitive message based on what the communicator believes to be in the audience's best interest. This approach is reflective of the dominance illustrated in Fig. 1.6.

Benefits of Culture-Centered Approach

In contrast to the culturally-sensitive approach, Dutta's (2007) culture-centered approach actually changes the social structures surrounding crisis and risk situations by involving underrepresented groups in the process of developing messages and communicating them to members of their respective cultural groups. Rather than relying on the risk communicator to determine what is in the best interests of the group, members of the group involved in the discussion of the situation provide their own perspectives of the risk or crisis situation and communicate the message to their respective group. The systemic convergence enables multiple bodies of knowledge to find voice in the process of communicating risk and crisis.

That said, the need to rethink the traditional model of risk and crisis communication seems apparent. Under the traditional model (Fig. 3.2), a spokesperson presents the message to the public. When attempting to be culturally-sensitive, the single spokesperson may provide scripted messages to cultural agents in order to appeal to different groups. The reality of multiple publics complicates this model because of the cultural variables that may influence how the scripted message is perceived and acted upon.

In the revised model (Fig. 3.4), during the pre-crisis phase, relationships are established with cultural agents drawn from the diverse publics who will be part of the message development phase and subsequent transmission if the risk becomes a crisis and a response is needed. The single spokesperson can still be at the center of the crisis and will likely serve as the contact person for the cultural agents who are ultimately responsible for developing and presenting the crisis message in a meaningful way to members of their respective cultural groups. This alternative approach is audience-centered and responds to the needs of people to get information from those who seem more closely affiliated with them.

	Developed message 1 ←→ Cultural Agent 1 ←→ Specific public
Single	*Developed message 2 ←→ Cultural Agent 2 ←→ Specific public*
Spokesperson ←→	*Developed message 3 ←→ Cultural Agent 3 ←→ Specific public*
	Developed message 4 ←→ Cultural Agent 4 ←→ Specific public
	Developed message…←→ Cultural Agent …←→ Specific public

Fig. 3.4 Culture-centered multiple spokesperson model of crisis communication

The culture-centered approach to risk communication has additional benefits in that it gives voice to the fears, frustrations, and concerns of multiple publics about the risk situation. The approach is interactive, thereby enabling all publics to have the opportunity to add their knowledge to the discussion. The script is developed, meaning that the publics have the opportunity to determine the words and ways in which the message is presented to their constituencies. The culture-centered approach is nonlinear and addresses issues based upon where individuals and particular audiences are within the spheres of ethnocentricity. The dialogic nature of the process encourages individuals to participate in the process through what has come to be known as community-based participatory research (CBPR).

CBPR has emerged in the public health arena as a way to involve community members, organizational representatives, and researchers in all aspects of the research process. As Israel et al. (2001) note: "Partners contribute their expertise and share responsibilities and ownership to increase understanding of a given phenomenon" (p. 184). They identify numerous advantages of CBPR: the enhanced "relevance and use of the findings by all partners involved"; the commitment of "partners with diverse skills, knowledge, and expertise in addressing complex problems"; the improved quality and validity of research by "incorporating the local knowledge of the people involved"; the increased possibility of "overcoming distrust of research on the part of communities that have historically been 'subjects' of such research"; "the potential to link across the cultural differences that may exist between partners involved"; and the potential to provide "resources for communities involved" (p. 185).

Summary

Historically, scholars in general and risk and crisis scholars in particular have ignored domestic multiculturalism when developing messages, operating under the following assumptions: well-constructed messages appeal to a broad, homogenous audience; cultural groups are more similar than different; crisis messages can be constructed following an established pattern; and the best way to communicate about a risk or crisis involves the use of a single spokesperson. Each of these assumptions can be mitigated by the literature of intercultural communication.

There is no homogeneous audience; there are multiple publics. Cultural groups vary greatly with regard to language, perceptions about their place in society, normative beliefs and values, and world view. Equifinality establishes that there are many equivalent ways to construct crisis messages for different publics, and solutions that integrate cultural perspectives will be more effective in communicating crisis messages to diverse publics through multiple spokespeople. The implication of this advice for risk communication is powerful. Rather than using one approach exclusively when conveying messages to multiple publics, we advocate the use of multiple spokespersons who can work with the central authority to disseminate risk and crisis information quickly and meaningfully to culturally diverse stakeholders.

References

Barton, L. (1993). *Crisis in organizations: Managing and communicating in the heat of chaos*. Cincinnati, OH: South-Western College Publishing/Thomson Learning.

Benoit, W. L. (1995). *Accounts, excuses, and apologies: A theory of image restoration strategies*. Albany, NY: State University of New York Press.

Benoit, W. L. (1997). A critical analysis of US Air's image repair discourse. *Business Communication Quarterly, 23*, 177–186.

Chess, C. (2001). Organizational theory and stages of risk communication. *Risk Analysis, 21*, 179–188.

Condon, J. C., & Yousef, F. (1975). *An approach to intercultural communication*. Indianapolis, IN: Bobbs-Merrill.

Coombs, W. T. (1998). An analytical framework for crisis situations: Better responses from a better understanding of the situation. *Journal of Public Relations Research, 10*(3), 177–192.

Coombs, W. T. (1999). Information and compassion in crisis responses: A test of their effects. *Journal of Public Relations Research, 11*(2), 125–142.

Coombs, W. T. (2007). *Ongoing crisis communication: Planning managing, and responding* (2nd ed.). Thousand Oaks, CA: Sage.

Covello, V. T. (1992). Risk communication: An emerging area of health communication research. In S. A. Deetz (Ed.), *Communication Yearbook 15* (pp. 359–373). Newbury Park, CA: Sage.

Covello, V. T. (2003). Best practices in public health risk and crisis communication. *Journal of Health Communication Research, 8*, 5–8.

Dutta, M. J. (2007). Communicating about culture and health: Theorizing culture-centered and cultural sensitivity approaches. *Communication Theory, 17*(3), 304–328.

Fink, S. (1986). *Crisis management: Planning for the inevitable*. New York: American Management Association.

Fong, M. (2006). The nexus of language, communication, and culture. In L. A. Samovar, R. E. Porter, & E. R. McDaniel (Eds.), *Intercultural communication: A reader* (11th ed.), (pp. 214–221). Belmont, CA: Thomson, Wadsworth.

Gudykunst, W. B., & Ting-Toomey, S. (1988). *Culture and interpersonal communication*. Newbury Park: Sage.

Hall, E. T. (1976). *Beyond culture*. New York: Doubleday.

Harris, P. R., & Moran, R. T. (1991). *Managing cultural differences* (3rd ed.). Houston: Gulf.

Heath, R. L. (1997) *Strategic issues management*. Thousand Oaks, CA: Sage.

Heath, R. L. (2001). Learning best practices from experience and research. In R. L. Heath (Ed.), *Handbook of public relations* (pp. 441–444). Thousand Oaks, CA: Sage.

Hofstede, G. (1991). *Cultures and organizations: Software of the mind*. London: McGraw-Hill.

Israel, B. A., Schulz, A. J., Parker, E. A., & Becker, A. B. (2001). Community-based participatory research: Policy recommendations for promoting a partnership approach in health research. *Education for Health, 14*(2), 182–197.

Klopf, D. W. (1991). *Intercultural encounters: The fundamentals of intercultural communication.* Engelwood, CO: Morton Publishing.

Kluckhohn, F., & Strodtbeck, F. (1961). *Variations in value orientations.* Evanston, IL: Row, Peterson.

Leanna, C. E., Ahlbrant, R. S., & Murrell, A. J. (1992). The effects of employee involvement programs on unionized workers' attitudes, perceptions, and preferences in decision making. *The Academy of Management Journal, 4*, 581–592.

Leitch, S., & Neilson, D. (1997). Reframing public relations: New directions for theory and practice. *Australian Journal of Communication, 24*(2), 17–32.

Lindell, M. K., & Perry, R. W. (2004). *Communicating environmental risk in multiethnic communities.* Thousand Oaks, CA: Sage.

Littlefield, R. S., & Cowden, K. (2006, November 17). *Rethinking the single spokesperson model of crisis communication: Recognizing the need to address multiple publics.* Paper presented to the Public Relations Division of the National Communication Association, San Antonio, Texas.

McMahan, S., Witte, K., & Meyer, J. (1998). The perception of risk messages regarding electromagnetic fields: Extending the extended parallel process model to an unknown risk. *Health Communication, 10*, 247–260.

Neuliep, J. W. (2003). *Intercultural communication: A contextual approach* (2nd ed.). Boston: Houghton Mifflin.

Perelman, C., & Olbrechts-Tyteca, L. (1971). *The new rhetoric: A treatise on argumentation* (2nd printing). (J. Wilkinson & P. Weaver, Trans.). Notre Dame, IN: University Press. (Original work published in 1958).

Rogers, E. M. (2003). *The diffusion of innovations* (5th ed.). New York: Free Press.

Rowan, K. E. (1991). Goals, obstacles, and strategies in risk communication: A problem-solving approach to improving communication. *Journal of Applied Communication Research, 19*(4), 300–329.

Samovar, L. A., & Porter, R. E. (2001). *Communication between cultures* (4th ed.). Belmont, CA: Thomson Wadsworth.

Sarbaugh, L. E. (1979). *Intercultural communication.* Rochelle Park, NJ: Hayden Book Company, Inc.

Schwartz, S. H. (1992). Universals in the content and structure of values: Theoretical advances and empirical tests in twenty countries. In S. H. Schwartz (Ed.), *Advances in experimental social psychology, 25* (pp. 1–66). San Diego, CA: Academic Press.

Seeger, M. W., Sellnow, T. L., & Ulmer, R. R. (1998).Communication, organization, and crisis. In M. E. Roloff (Ed.), *Communication Yearbook, 21* (pp. 231–276). Thousand Oaks, CA: Sage.

Seeger, M. W., Sellnow, T. L., & Ulmer, R. R. (2003). *Communication and organizational crisis.* Westport, CT: Praeger.

Slovic, P. (1986). Informing and educating the public about risk. *Risk Analysis, 6*(4), 403–415.

Tierney, K. J. (1999). Toward a critical sociology of risk. *Sociological Forum, 14*(2), 215–242.

Ting-Toomey, S. (1989). Language, communication, and culture. In S. Ting-Toomey & F. Korzenny (Eds.), *Language, communication, and culture* (pp. 9–15). Newbury Park, CA: Sage.

Ting-Toomey, S., & Chung, L. C. (2005). *Understanding intercultural communication.* Los Angeles: Roxbury.

Weick, K. E. (1988). Enacting sensemaking in crisis situations. *Journal of Management Studies, 25*(4), 305–317.

Weick, K. E. (1995). *Sensemaking in organizations.* Thousand Oaks, CA: Sage.

Part II
Cases in Risk Communication

Chapter 4
The Case Study Approach

"The person who can combine frames of reference and draw connections between ostensibly unrelated points of view is likely to be the one who makes the creative breakthrough."

—Denise Shekerjian

In the previous section, our framework describing the interaction of multiple competing messages provided a useful way to describe how risk communicators should create convergence and understanding with their audiences in pre-crisis, crisis, and post-crisis situations. In addition, the identification of best practices offers a way to identify why particular risk messages may have more influence than others on how audiences respond. Adding to the complexity of the situation for risk communicators are multiple publics who may not share the same understanding or willingness to respond to the messages due to how the risk or potential crisis may affect them in what we described as spheres of ethnocentricity.

In the complex communication context of risk communication, one research methodology is particularly appropriate, due to its capacity to explore, describe, or explain the dynamics of the situation. The case study approach to research in the social sciences is a fitting method for identifying the interaction between individuals, messages, and context. Yin (2003) summarizes, "The case study method allows investigators to retain the holistic and meaningful characteristics of real-life events" (p. 2). The case study approach works well to identify best practices for risk communication because individual situations are defined or isolated, relevant data are collected about the situation, and the findings are presented in such a way that a more complete understanding is reached regarding how messages shape perceptions and serve to prompt particular responses from those hearing the messages.

We consider the case study method as both an approach to research and a choice of what to study (Patton, 2002). Therefore, in the construction of the case studies presented in this book, a consistent methodological approach was followed. In order to establish common areas of analysis within the research design, we focused on risk situations involving the unintentional or intentional contamination or compromise of the food system. A conceptual framework based upon the best practices explained in Chapter 2 and a chronological exposition of the pre-crisis, crisis, and post-crisis messages created consistency as we drew implications about what happened, how it happened, and why. Collectively, the cases allowed us to generalize about the best practices as a whole.

T. L. Sellnow et al. *Effective Risk Communication*

DOI: 10.1007/978-0-387-79727-4_4

Individually, the choice of cases provided opportunities to demonstrate various aspects of the best practices. Each of the forthcoming cases includes particular situations and context-sensitive information whereby the best practices for risk communication could be identified and studied. In some cases, preemptive communication strategies designed to promote compliant behavior are found. In others, the exposition of the crisis revealed how risks were not anticipated or communicated effectively to the public. In fact, as demonstrated by how a crisis actually unfolded, the evidence suggests that those managing the crisis were not always mindfully considering the competing arguments. Rather than seeking congruence, reliance on economic or social models enabled decision-makers to simplify the risk situation at a time when complexity should have been acknowledged. In such situations, the value of the best practices may not have been recognized until after the crisis had passed.

Justification for the Case Study Approach

We selected the case study approach as a way to illustrate the interactive process involved in the convergence of risk messages for several reasons. Case studies have been used frequently by scholars and practitioners in public health, agriculture, education, psychology, and the social sciences as a legitimate methodological approach to research (Rogers, 2003; Tuschman & Anderson, 1997). In addition, they provide a method to investigate a contemporary event involving risk within a real life context; and they contribute to enhanced knowledge of complex social phenomena.

Legitimacy as a Methodological Approach

The case study approach has been used to study many different situations involving individual, group, organizational, social, political, and related phenomena (Yin, 2003). Throughout his treatise on diffusion theory, Rogers (2003) offers cases to illustrate the following: social systems (Iowa hybrid corn case), pro-innovation bias (Egyptian villages pure drinking water case), socioeconomic status (California hard tomatoes case), the reinvention process (horse culture among the Plains Indians case), attributions of innovations (photovoltaics, cellular telephones case), adopter types (Old Order Amish case), opinion leadership (Alpha Pups in the viral marketing of an electronics game case), diffusion networks (London Cholera epidemic case), change agents (Baltimore needle-exchange project case), stages in the innovation process (Santa Monica freeway diamond lane experiment case), and the consequences of innovations (steel axes for stone-age Aborigines case), to name but a few.

In their collection of readings on managing strategic innovation and change, Tuschman and Anderson (1997) offer numerous case studies involving technology cycles, design changes, power dynamics in organizations, managing research and development, product development, cross-functional linkages, and leadership styles. Similarly, risk and crisis scholars have used case studies to illustrate best practices and organizational learning.

Sellnow and Littlefield (2005) use case studies describing both accidental and intentional contamination to demonstrate lessons learned about protecting America's food supply. Three cases focused on particular companies and their experiences managing a crisis: Schwan's demonstration of social responsibility in response to a *Salmonella* contamination crisis, Chi-Chi's inability to survive a Hepatitis A outbreak despite apologia, and Jack in the Box restaurants' organizational learning following an *E. coli* outbreak in Seattle, Washington. A case involving interagency coordination and the tainted strawberries in the National School Lunch Program revealed how various stakeholders affect crisis planning efforts. Two cases of potentially intentional contamination–one by Monsanto, a major producer of genetically engineered wheat and the other by the Boghwan Shree Rajneesh cult in an Oregon community–explored the effect of public opinion and outrage.

In their work, Ulmer et al. (2007) provide case studies revealing lessons learned about managing uncertainty, effective communication, and demonstrating leadership. They focused on four areas: industrial disasters (Exxon and the *Valdez* oil tanker, and the fires at Malden Mills and Cole Hardwoods), food borne illness (Jack in the Box's *E. coli* O157:H7, Hepatitis A at a Chi-Chi's restaurant, and the Schwan's *Salmonella* crisis), terrorism (the case of 9/11, the Oklahoma City bombing, and the CDC's handling of the SARS outbreak), and natural disasters (the 1997 Red River Valley floods, the Tsunami and the Red Cross, and the 2003 San Diego County fires).

Exploring risk and crisis situations in public health, Seeger et al. (2008) categorize case studies focusing on bioterrorism, food borne illness, infectious disease outbreaks, and crisis prevention and responses. Cases of bioterrorism focused on lessons learned from the 2001 Anthrax crisis through the U.S. Postal System; the threat of agro-terrorism in high reliability organizations and organizational responses to the Chi-Chi's Hepatitis A outbreak provided cases demonstrating the risk of food-borne illness and the need for crisis prevention; and the strategies used by communities, nations, and the world when dealing with the risk of West Nile Virus, SARS, Encephalitis, HIV and AIDS provided cases where infectious disease outbreaks required effective risk and crisis responses. These collections and others similarly have found value in studying particular examples of an identified phenomenon for the benefit of understanding more about what, how, and why something happened.

Multiple Sources of Information

One of the reasons supporting the legitimacy of the case study approach is its use of multiple sources of information to establish claims about a particular situation.

Type	Examples
Textual Materials	National Newspapers *New York Times, Wall Street Journal* Regional Newspapers *Boston Globe* International Newspapers *The Financial Times, The Dominion Post*
On-line Materials and Resources	Government Websites U.S. Department of Health and Human Services Department for the Environment, Food and Rural Affairs Industry Websites Odwalla, Inc. Kidsource Online ConAgra Foods
Interviews and Official Statements	Telephone interviews On-line interviews Official statements and press releases
Media Accounts	Domestic and international television coverage Public and private radio coverage

Fig. 4.1 Multiple sources of information used in case studies

Multiple sources may include textual materials, on-line websites and resources, interviews, media accounts, and personal observations. Due to the nature of the case study approach, choices must be made about the kinds of information to be utilized. Accessibility often dictates the kinds of information to be included, in which case the researchers must continually cross reference to be sure that the most accurate depiction of the situation is conveyed.

For the case studies included within this volume, text-based materials provided the majority of the information consulted (Fig. 4.1). Information drawn from national newspapers (e.g., *New York Times* and *The Wall Street Journal*), regional newspapers (e.g., *The Boston* Globe) or–as in the New Zealand foot and mouth hoax case–international outlets (e.g., *Financial Times* and *The Dominion Post*) provided contextual material enabling the researcher to establish the time frame and variables at work in each case. Websites and on-line materials, such as those offered by governmental and industrial groups provided insight from the perspective of those in positions to respond to the risk or crisis situation. A number of groups were accessed through such websites, including the U.S. Department of Health and Human Services, the Department for the Environment, Food, and Rural Affairs, Odwalla, Inc., KidSource Online, and ConAgra Foods. Interviews were conducted with individuals holding positions of responsibility, enhancing the researcher's understanding of the dynamics of the situation in New Zealand.

When interviewing was impossible, official comments from key decision-makers were drawn from the available textual sources. Together, these multiple sources enabled the observer to engage in triangulation, a process where more than one source of information is used when drawing inferences or conclusions about a given situation. Stake (2000) argues that triangulation was valuable not only to clarify meaning, but also to identify "different ways the phenomenon is being seen" (p. 444).

Need for Theoretical Framework

In addition to the need for multiple sources of data to understand the complexity of a risk or crisis situation, another reason researchers use the case study approach stems from the way theoretical propositions may be used to guide data collection and analysis. Case study researchers can set the parameters for what will be included within the analysis. As such, the introduction of a theoretical framework provides an overlay for the data that the researcher may use as a way to explore, describe, or explain what happened. In the selected cases included in this volume, the researchers utilized existing theoretical perspectives about risk and crisis drawn from the professional journals of the field, including *Journal of Applied Communication Research, Management Communication Quarterly, Journal of Epidemiol Community Health*, and *The New England Journal of Medicine*. The existing theoretical framework provided a backdrop for considering each case.

Specifically for this volume, best practices for risk communication were used as a theoretical framework (Fig. 4.2). As already explained in Chapter 2, these best practices are theory driven and stem from the work previously done through a collaboration of risk and crisis communicators who introduced the ten best practices for crisis communication through the National Center for Food Protection and Defense (Seeger, 2006). Each of the case studies used these best practices to help to reveal problems faced by risk and crisis communicators, as well as to identify the strategies used as individuals, organizations, and communities worked to move through the crisis to recovery and in some cases, renewal.

1. *Infuse risk communication in policy making.* As organizations or agencies establish policies, risk communication may reiterate previous arguments, consider current arguments, and/or introduce new arguments.
2. *Treat risk communication as a process.* The dynamic nature of risk situations requires managers and communicators to continuously review competing arguments in the construction of risk messages.
3. *Account for the uncertainty inherent in risk.* Risk communicators must acknowledge and reinforce the unknown as an argument when framing messages for the public.
4. *Design risk messages to be culture-centered.* To achieve the desired response, risk communicators must work with diverse publics to develop messages that are meaningful and efficacious.
5. *Acknowledge diverse levels of risk tolerance.* Due to widely varying capacities to process risk messages due to perceptions of hazard and outrage, communicators must recognize complexity and capacity when constructing risk messages.
6. *Involve the public in dialogue about risk.* The public's right to know about potential risks, and their role as participants in finding interacting arguments may prompt less hazard and outrage.
7. *Present risk messages with honesty.* Inherent in the process of finding congruence is the recognition that arguments are presented truthfully and completely.
8. *Meet the risk perception needs by remaining open and accessible to the public.* In the process of finding convergence, the public must have access to those creating risk messages for clarification and assurance.
9. *Collaborate and coordinate about risk with credible information sources.* For messages to be considered credible, groups and agencies must interact with each other about risk situations and share information.

Fig. 4.2 Best practices of risk communication

Utility for Investigation into Contemporary Events

In addition to the case study method being multidimensional, researchers value the approach because it provides an empirical way to investigate a contemporary phenomenon within a real life context. There are differences between the case study approach and studies utilizing a more structured research methodology. For example, in an experimental setting, variables may be controlled or accounted for as particular actions are taken to affect the outcome. In this closed environment, researchers can make generalizations based upon the sophistication of their design.

However, situations where risk messages are communicated through the media and events are reported and presented as they unfold, researchers have less control over how competing risk message are transmitted and received by diverse groups within the public. The range of variables that cannot be controlled or manipulated further complicates the coverage of contemporary events outside of the laboratory. For example, Chapter 8 examines the case of in the tainted Odwalla juice, the *Cryptosporidium* outbreak case, or in the case of finding *Salmonella* in ConAgra Foods pot pies, human error could not have been predicted with certainty. In disasters like Hurricane Katrina, elements of nature could not be controlled. They happened. In the New Zealand hoax case, the potential threat of a terrorist's intentional foot and mouth disease could not have been precluded. The realization that a host of variables are interacting in a real-world setting affords the scholar a unique opportunity to explore, describe, and explain events as they occur.

The opportunity to examine what transpired in a particular crisis situation is unique to the case study approach. Due to the dynamic, chaotic nature of crisis events that are not always represented in cause-to-effect relationships, the case study approach enabled us to examine and understand situations in ways that might not have been foreseen prior to the start of our investigation. While not statistically generalizable, after examining several cases, the identification of the presence or absence of the best practices provides researchers with the arguments needed to find consistency about the situation that may have applicability to other similar risk situations. By using the case study framework to separate the pre-crisis from the crisis, observers may note events leading up to the crisis, factors that may have contributed to the way the risks were presented, and what happened (or should have happened) as a result of the way these messages were processed and acted upon.

Enhancement of Knowledge About Complex Phenomena

Within any given situation involving risk, there are many variables of interest, including the processes at work within the dynamic of the situation; the changes that occur due to the introduction of particular risk messages; relations between various stakeholders during the pre-crisis, crisis, or post-crisis situations; and the learning that results following the response to a crisis situation. These variables involving individuals, groups, organizations, or social entities represent the multidimensionality of the phenomena involved in risk communication.

In addition, the varied nature of questions posed by researchers and practitioners pertaining to risk situations further demonstrates how the complexity of a case can be studied using this approach. Case studies seeking to know what happened in a particular context rely on "what" questions. What happened in a particular situation causing a response? When researchers seek answers to "how" questions, they want descriptions. How did an entity communicate risk messages? Researchers seeking explanations regarding the particular motivations of communicators in a situation rely on "why" questions. Why were company spokespeople compelled to communicate particular messages to the public about a risk situation? In contrast to quantitative and qualitative methodologies where researchers tend to focus on one dimension or variable, the case study approach enables the researcher to use all of these questions. Questions like these are used throughout the cases to reveal the multidimensionality of the risk and crisis events.

Establishing a Framework for Case Studies

Identifying a framework for case studies is essential if comparisons are to be made. Stake (2000) suggests the following items as essential in the creation of a case study:

> The nature of the case; the case's historical background; the physical setting; other contexts (e.g., economic, political, legal, and aesthetic); other cases through which this case is recognized; and those informants through whom the case can be known. (pp. 438–439)

Thus, to provide clarity, we determined that each case study should be written to include common elements providing comparable information for the reader to consider. In the following case studies, we provide:

- An introduction and overview of the case.
- Evidence and application of the best practices for risk communication within the case.
- Lessons learned and implications drawn from the use of best practices for risk communication.

From a research perspective, with these elements as constants, individual authors were able to gather data appropriate to each case and uniformly present their findings. In addition, similar textual materials were used in each of the studies, providing the reader with comparable information to consider.

Five Cases of Risk Communication

Stake (2000) argues that, "perhaps the most unique aspect of the case study is the selection of cases to study" (p. 446). With this in mind, we selected crisis situations where the presence or absence of best practices for risk communication could be identified, providing the readers with insight into how risk communication may or could have been used to affect the behavior of various stakeholders prior to the onset of a crisis situation. Each case study is unique in the risks posed, as well as how the communication agent sought to affect compliant behavior from the various stakeholders receiving the risk messages.

In the case of the *Cryptosporidium* crisis, the risks associated with water quality in a major metropolitan area and a community's response to a water quality crisis are examined. The risks associated with inadequate planning and the events related to the Hurricane Katrina disaster reveal how different levels of the government responded to a natural disaster. A government's use of interacting arguments revealed a paradox between appearing to accept the risk of foot and mouth disease and dismissing its likelihood on a New Zealand island. The Odwalla case study focuses on the risks associated with their trademark apple juice and how it struggled to renew itself within the health food industry. Finally, the ConAgra Foods *Salmonella* case study features the complexities of addressing multiple audiences during a major recall event involving pot pies.

"*Cryptosporidium: Unanticipated Risk Factors*," provides the example of a community organization that experienced a crisis because it did not respond in time to government warnings calling for stronger guidelines for guarding municipal water against *Cryptosporidium* invasions. At the time of the crisis, Milwaukee had no water monitoring systems in place and the outbreak served as a wake-up call by exposing weaknesses in the public health system and pointing out the bioterrorism risks. Throughout the crisis, community leaders failed to be open, honest, and timely with the information they provided to the public. They failed to be mindful of public concerns expressed prior to the cryptosporidium outbreak. In addition, they did not collaborate or coordinate across agencies, exacerbating the crisis. Milwaukee was unprepared but learned from the event, established a plan should such a crisis occur in the future, and now has one of the safest water treatment systems in the country.

"Hurricane Katrina: Risk Communication in Response to a Natural Disaster," examines how local leaders failed to create an adequate crisis plan, despite having knowledge of the damage that would occur if a hurricane of Katrina's magnitude struck New Orleans. While local crisis managers had a plan, its usefulness was mitigated by the length and format of the document. Once the hurricane struck, New Orleans crisis managers faced the difficult challenge of collaborating and coordinating resource distribution to affected residents. Another difficulty was getting information to the stakeholders. In the pre-crisis and crisis stages, the media were often ahead of local officials in presenting information to residents. This compromised the local officials' credibility and accountability. Clearly, lack of pre-event planning, the absence of collaboration and coordination, and the need for honest, candid, open, and accountable communication are key reasons why local crisis managers were unable to plan for, manage, and move past what was a devastating event for New Orleans and the surrounding region.

"New Zealand Beef Industry: Risk Communication in Response to a Terrorist Hoax" expands knowledge of risk communication by introducing how hoaxes and terrorist threats complicate our understanding of risk situations. In New Zealand, after receiving a threat claiming a deliberate release of the foot and mouth disease virus on Waiheke Island, the government had to provide intersecting messages demonstrating their capacity to manage the crisis situation. In essence, they claimed to be treating the situation as a potential crisis, while at the same time indicating their belief that the threat was a hoax. Managing crisis uncertainty became the focus for local leaders as they presented messages minimizing the risk as a hoax while acknowledging their treatment of the message as a viable threat to the security of the cattle and the New Zealand economy. The pre-crisis partnerships established between crisis managers and the various stakeholders proved valuable as the agencies worked together to disseminate information and communicate with the local citizens, as well as New Zealand's international partners. While a crisis plan was in place, and had been tested, there were some initial concerns raised by the local public that were ultimately mitigated due to open communication, as well as an attitude of compassion and empathy demonstrated by crisis spokespeople. While the hoax never developed into a crisis, providing messages of self-efficacy about checking for symptoms became an effective way to garner public confidence.

How a company managed to survive the challenge of an *E. coli* outbreak associated with one of its juice products is the subject of the chapter, "Odwalla: The Long Term Implications of Risk Communication." Despite the potential risks associated with the continued consumption product, the public stood by Odwalla and its actions during and after the crisis. Odwalla met the needs of the media and remained accessible by holding press conferences, continually updating a website, instituting a hotline, and maintaining open communication with consumers and the press. The company delivered messages of self-efficacy and offered multiple ways for consumers to remain safe. In addition, company leaders apologized publicly, acknowledged the tragedy of the situation, paid medical bills for victims, and acknowledged the impact of the crisis on the image of the company. Following the crisis, Odwalla created an advisory council that ultimately recommended a new pasteurization process, breaking new ground in the industry. The use of some of the best practices enabled Odwalla to embrace a crisis, use it as an opportunity to become an industry leader, initiate industry wide change, and to encourage organizational renewal.

"ConAgra: Audience Complexity in Risk Communication" focuses on the need for organizations to consider multiple audiences when issuing risk messages. In the process of what appeared to be a demonstration of more concerned about their bottom line than with the safety of their customers, ConAgra initially shifted the blame for the outbreak to consumers for not cooking the pot pies properly. In addition, ConAgra made overly-assuring statements to the public about which products were affected by *Salmonella* (chicken and turkey), and which were not (beef). The assumptions made by ConAgra Foods about the literacy levels, economic status, access to media, proximity to outbreak, and cultural group identities of those receiving the risk messages also complicated their communication with stakeholders. In this case, once *Salmonella* was linked to ConAgra Foods' pot pies, the company issued a recall of all brands associated with their product. While additional information about the ConAgra Foods recall has yet to emerge, the case points to the need for greater attention by company spokespeople to the best practices of risk communication in order to preserve a positive reputation with the public.

Summary

This chapter introduced the case study method as a viable way to study risk communication in crisis situations. Our reasons for choosing the case study approach include its utility for exploring situations from multiple points of view, its usefulness when investigating contemporary events, and its ability to provide enhanced knowledge about complex phenomena. The best practices of risk communication, based on the best practices of crisis communication (Seeger, 2006), provide the theoretical framework for the case studies included in this book.

The framework we used for the case studies includes an introduction and overview to the case, a timeline of events, evidence and application of the best practices, and lessons learned. Five cases were introduced: the Milwaukee *Cryptosporidium* crisis, the Hurricane Katrina crisis, the New Zealand foot and mouth disease hoax crisis, the Odwalla juice crisis, and the ConAgra Foods *Salmonella* crisis. In each case, the best practices of risk communication provide insight into what occurred, or failed to occur, and the implications that followed in each crisis situation.

The five case studies provide insight into the best practices of risk communication. In all of the cases, risk communicators should have acknowledged competing arguments in the construction of risk messages. For example, the *Cryptosporidium* case demonstrates the need to infuse risk communication into policy making. By accepting that current practices would take care of the problem, local leaders allowed the crisis to develop. In the Hurricane Katrina case, the dynamic state of affairs required communicators to continuously review the situation and be proactive in communicating strategies of self-efficacy. Regarding the New Zealand potential foot and mouth disease case, a clear argument exists for why risk communicators must acknowledge and reinforce the unknown when framing messages for the public. Similarly, the collaboration and coordination among agencies with credible information sources helped the New Zealand crisis leaders build support among the various stakeholders affected by the potential contamination. As for Odwalla, the company was forced to acknowledge diverse levels of risk tolerance as the complexity of the situation unfolded. Similarly, a recognition of the need for a culture-centered approach would have enhanced the communication of ConAgra Foods with consumers and demonstrated a commitment to safety over profit. The following five chapters serve as examples of case studies involving risk communication.

References

Patton, M. Q. (2002). *Qualitative research and evaluation* methods (3rd ed.). Thousand Oaks, CA: Sage.

Rogers, E. M. (2003). *Diffusion of innovations* (5th ed.). New York, NY: Free Press.

Seeger, M. W. (2006). Best practices in crisis communication: An expert panel process. *Journal of Applied Communication Research, 34*(3), 232–244.

Seeger, M. W., Sellnow, T. L., & Ulmer, R. R. (Eds.). (2008). *Crisis communication and the public health*. Cresskill, NJ: Hampton Press, Inc.

Sellnow, T. L., & Littlefield, R. S. (Eds.). (2005). *Lessons learned about protecting America's food supply: Case studies in crisis communication*. Fargo, ND: Institute for Regional Studies.

Stake, R. E. (2000). Case studies. In N. K. Denzin & Y. S. Lincoln (Eds.), *Handbook of qualitative research* (2nd ed.), (pp. 435–454). Thousand Oaks, CA: Sage.

Tuschman, M. L., & Anderson, P. (1997). *Managing strategic innovation and change: A collection of readings*. New York: Oxford University Press.

Ulmer, R. R., Sellnow, T. L., & Seeger, M. W. (2007). *Effective crisis communication: Moving from crisis to opportunity*. Thousand Oaks, CA: Sage.

Yin, R. K. (2003). *Case study research: Design and methods* (3rd ed.). Applied Social Research Methods Series, Vol. 5. Thousand Oaks, CA: Sage.

Chapter 5
Cryptosporidium: Unanticipated Risk Factors

Written with Devon Wood

In April 1993, Milwaukee experienced the largest outbreak of waterborne Cryptosporidiosis in U.S. history. The official toll was "403,000 sickened, 44,000 doctor visits, 4,400 hospitalized, more than 100 deaths, 725,000 lost work or school days, $96 million in lost wages and medical expenses and $90 million for a new water purification system" (Marchione, 2003, p. 01B). The outbreak caught the city off guard because Milwaukee Water Works (MWW) officials and the Milwaukee Health Department had inadequate policies in place to monitor city water. Consequently, this case became an important learning opportunity for the Centers for Disease Control and Prevention (CDC), the Milwaukee Health Department, and city health officials nationally.

This case provides an example of what happens when a city fails to adequately manage risks associated with its water treatment facility. Ultimately lessons learned from its crisis led to improvements for Milwaukee and the entire water treatment industry. This case proceeds as follows: (1) an overview of the case, including a time line, is given; (2) the case is then analyzed using the best practices in risk communication; and finally (3) practical implications for effective risk communication identified from this case are provided.

Managing Risk Communication in Milwaukee's Water Treatment Facility

Before the Milwaukee crisis, there were few regulations for monitoring public water supplies. In fact, after the crisis an article in the *New England Journal of Medicine* concluded that "water quality standards and the testing of patients for Cryptosporidiosis were not adequate to detect this outbreak" (MacKenzie et al., 2004, p. 161). MacKenzie et al. (1994) provide useful information about how the risk increased in the Milwaukee Water Works (MWW) plant incrementally: "At the time of the outbreak, both MWW plants treated water by adding chlorine, polyaluminum

T. L. Sellnow et al. *Effective Risk Communication*

DOI: 10.1007/978-0-387-79727-4_5

chloride ... and filtration particles to clean the water" (p. 161). An examination of plant records showed there were increased levels of water turbidity (water cloudiness) around the time of the event (MacKenzie et al., 1994). In addition they describe that key monitoring processes enabled the risk to increase without identification. They explain:

> Inspection of the southern plant revealed that a streaming-current monitor, which can aid plant operators in adjusting the dose of coagulant, had been incorrectly installed and thus was not in use. In addition, monitors designed for continuous measurements of the turbidity of filtered water were not in operation. Turbidity was monitored about once every eight hours. (MacKenzie et al. 1994, pp. 163–164)

When residents complained about "cloudy, foul-smelling water ... city health officials kept using more of the chemical to try to fix the problem, not realizing it was ineffective" (Marchione, 2003, Para. 21). At the time of the crisis, the water "met all federal and state standards" (Marchione, 2003, Para. 3), in spite of the fact that "government agencies received warnings that stronger guidelines were needed for guarding municipal water against invasions by *Cryptosporidium*" ("EPA: Ignored Warnings," 1993, p. 9A).

Although there were signs of a problem prior to the crisis, city officials failed to make the connection between the water supply and the fact that an unusually large number of people were becoming ill with similar symptoms. By the time city officials realized there was a connection, it was too late. Had communication between agencies been stronger, the connection could have been made sooner.

Eventually, when a patient tested positive for the parasite *Cryptosporidium*, leading city officials made the connection. Marilynn Marchione (2003), a medical correspondent for the Milwaukee *Journal Sentinel* reported:

> In a meeting to discuss what to do, the mayor [John O. Norquist] set a glass of water in front of Davis, the state's chief medical officer, and asked whether he'd drink it. Davis said no. Norquist called a 9 p.m. news conference and ordered residents in Milwaukee and 10 suburbs that use city water to boil it until further notice. (p. 3)

Other efforts to inform the public were made by local newspapers: "The *Milwaukee Journal* and the *Milwaukee Sentinel* abandoned competitive publishing and jointly published a special edition in Spanish and English, warning people not to drink the water" (Marchione, 2003, p. 3). Just after the boil water advisory was lifted, Paul Nannis, Milwaukee's Health Commissioner, announced, "Some food products made with Milwaukee tap water could be recalled nationwide. Federal and local health officials are surveying local food manufactures to compile a list" ("Its water," 1993, p. 16). Nannis also reported that "city officials told residents to run their tap water for three minutes to flush pipes of any stagnant water" (p. 16). These efforts may have helped, but unfortunately the contaminated water had already reached many homes and businesses. People were frustrated because of officials' slow response. When the public needed risk communication they received none. No one was available to answer or return residents' calls. In fact, city officials avoided dealing with the problem simply because they did not know enough about it.

Milwaukee's waterborne Cryptosporidiosis epidemic led to fundamental changes nationally in public health, surveillance, and water treatment. Together with the CDC, the U.S. Department of Health and Human Services developed specific guidelines and protocols for dealing with water-borne illness outbreaks. In addition, an emphasis was placed on the need for open communication networks with common stakeholders such as public health officials and physicians. The city of Milwaukee developed its own laboratory for studying the parasite *Cryptosporidium* and now has some of the most protected water in the nation. This outbreak served as a wake-up call because it exposed flaws in the public health system. From this experience, lessons learned have helped the public health system mature in its preparedness for bioterrorism and water safety (Fig. 5.1).

Mid-to-late 1992	Milwaukee Water Works officials switch purification chemicals, going from alum to polyaluminum chloride.
March 1993	Spring snow melt increases runoff into Milwaukee River flowing into Lake Michigan near an intake pipe for the Howard Avenue treatment plant.
March 23-April 9	City's Howard Avenue plant notes marked increases in treated water turbidity (1.0 to 1.7). Abnormally high, but well within state standards.
March 29	Customer complaints reach a peak. Local health department notices an unusually high number of absences in schools and hospitals. Drugstores sold out of anti-diarrhea medications.
April 2	After experimenting with various amounts of polyaluminum chloride, Water Works officials switch back to alum.
April 5	Health Department receives calls from citizens regarding stomach complaints. Hospitals have unusually high numbers of patients with gastrointestinal illness. Milwaukee Department of Health contacts Wisconsin Division of Health.
April 6	Patient tests positive for *Cryptosporidium*. Health Department officials order other tests.
April 7	Tests on stool samples indicate presence of *Cryptosporidium*. Mayor John Norquist issues a boil order after the state epidemiologist declines to drink a glass of water.
April 9–11	Telephone survey conducted to determine representative clinical characteristics of illness among the people affected.
April 10	Howard Avenue treatment plant chemist Eugene Marks admits plant managers failed to recognize or respond to warning signs and says the plant could have been shut down when turbidity reached higher than normal levels.
April 13	Tap water leaving the Howard Avenue plant in late March and early April reported to be the most turbid in years. For six days, cloudiness at the Howard Avenue plant topped 1.0.
April 14	Public Works official James Kaminski acknowledged no policies or procedures in place to notify public health officials about turbidity problems.
April 15	Norquist lifts the boil order after tests show the water is again safe.

Fig. 5.1 Timeline

Applying the Best Practices of Risk Communication to the Milwaukee *Cryptosporidium* Outbreak

Cryptosporidium is a water-borne parasite that spreads easily throughout a water supply. In a CDC *Morbidity and Mortality* weekly report, the U.S. Department of Health and Human Services (1995) explains that a *Cryptosporidium* "infection can be transmitted through person-to-person or animal-to-animal contact, ingestion of focally-contaminated water or food, or contact with focally-contaminated environmental surfaces" (p. 2). Knowing this only adds to the uncertainty surrounding the parasite. However, *Cryptosporidium* has existed for years and it is not uncommon for the parasite to be found in municipal water supplies.

Corso et al. (2003) define *Cryptosporidium* as "a protozoan parasite that causes gastrointestinal illness" that "is transmitted by ingestion of oocysts excreted in human or animal feces" (p. 426). Cryptosporidiosis especially affects those with weakened immune systems but can cause severe gastrointestinal illness in people with healthy immune systems. Mac Kenzie et al. (1994) argue that "*[C]ryptosporidium* has been recognized as a cause of... illness in both immunocompetent and immunodeficient people.... In immunocompentent people, Cryptosporidiosis is a self-limited illness, but in those who are immunocompromised, infection can be unrelenting and fatal" (p. 161).

The CDC (1995) further highlights that a "high priority should be placed on educating immunosuppressed persons, who are at increased risk for severe Cryptosporidiosis if they become infected (p. 10). Laura Beil (1995), a reporter for the *Dallas Morning News* said, "About a week after becoming infected, a person begins to experience diarrhea, fever, vomiting, stomach cramps and other symptoms" (p. F8). These symptoms can and have been fatal for the elderly and those with immune complications.

Milwaukee officials' handling of the risk associated with *Cryptosporidium* had implications, both positive and negative, for the overall outcome of this epidemic.

Account for the Uncertainty Inherent in Risk

At the time of the crisis, MWW did not have proper monitoring systems in place. Looking back on the series of events leading up to the boil water advisory, it is apparent that Milwaukee city officials did not believe the threat was as serious as it turned out to be. According to Ulmer et al. (2007), while "some people in the organization may view a situation as a potential crisis and others may not... communication about potential threats helps reduce uncertainty about potential risks in the organization" (p. 20). The MWW never discussed *Cryptosporidium* as a potential threat to the water supply.

An exact source for the contamination was never determined, but after gaining a better understanding of the parasite, it became obvious what caused the outbreak. It is now understood that chlorine does not kill *Cryptosporidium*, but that was not thought to be the case at MWW. In an article written 10 years after the event, Marchione (2003) reports that there were early signs of problems: "For weeks, residents called the water department, complaining of cloudy, foul-smelling water, and officials kept using more of the chemical [chlorine] to try to fix the problem, not realizing it was ineffective" (p. 01B). A variety of other signs leading up to the event were missed:

> There were unusual weather conditions... that affected currents and how water flowed around the intake pipe in Lake Michigan, the source of the city's water. City officials also discovered an illegal sewer connection that had been allowing waste from a slaughterhouse to enter sanitary sewers. "It was an unbelievable confluence of events. There were so many things going on at the same time," said state epidemiologist Jeff Davis. By April 5, a Monday, calls were flooding the Milwaukee Health Department from the public, the media and pharmacies saying people were sick, and diarrhea medications were flying off shelves.... For several days, officials tested for bacteria and viruses and came up blank. The protozone didn't come to light until the afternoon of April 7, when community doctor Thomas Taft called Health Department to say that another doctor, Anthony Ziebert, had an older patient who tested positive for *Cryptosporidium*. (Marchione, 2003, p. 01B)

Later in April, the city determined that *Cryptosporidium* had been pumped into the water from Lake Michigan. Brian Buggy, an infectious disease specialist at a Milwaukee hospital, said, "By the second week of the epidemic, we knew that we knew that we didn't know" ("EPA: Ignored Warnings," 1993, p. 9A). Anti-diarrhea medicine sales also indicated a problem:

> At a Cub Foods supermarket, the first sign was a sudden shortage of Kaopectate and Pepto Bismol. "We couldn't keep them on the shelves," said store manager Dennis Lipofski. The same was true all over. "We couldn't get more products to replace them." ("Milwaukee seeks source," 1993, Para. 2)

Surprisingly enough, the outbreak happened even though water met all federal and state standards. It happened without health officials being aware of it for many days. Five months after the epidemic, the *Capital Times* reported:

> Walter Jackubowski, EPA parasite specialist, said "procedures used to detect and count *Cryptosporidium* in water samples were cumbersome, costly and not accurate at the time." ... [and in addition to that] Robert Baumeister, chief of public water programs for the state Department of Natural Resources, said states are busy complying with many EPA directives and spend little time on exceeding EPA's guidance. ("EPA: Ignored Warnings," 1993, p. 9A)

The Cryptosporidiosis outbreak "exposed a potentially serious weakness in the US system of protecting drinking water" (Allen, 1995, p. 17). It was found that "for two decades, government agencies have received warnings that stronger guidelines are needed for guarding municipal water against invasions by *Cryptosporidium*" ("EPA: Ignored Warnings," 1993, p. 9A). It is clear from best practices in risk communication that planning is instrumental in reducing the uncertainty surrounding risk. Seeger (2006) explains:

> Planning has a variety of benefits. These include identifying risk areas and corresponding risk reduction, pre-setting initial crisis responses so that decision making during a crisis is more efficient, and identifying necessary resources. (p. 237)

Clearly, the Milwaukee city officials did not prepare or manage the risk uncertainty of this event effectively. They were caught off-guard and actually added to the uncertainty of the event. What follows are four key issues that augmented the risk uncertainty surrounding this event:

(1) Initial signs were not perceived to pose a threat.
(2) The signs were not handled with the immediacy they should have been.
(3) Officials were not communicating effectively about the need to better monitor the system
(4) MWW did not consider and communicate about potential threats. Not having proper systems and practices in place set the Milwaukee city officials up for disaster during their crisis response.

Collaborate and Coordinate about Risk with Credible Information Sources

A key element of risk communication discussed in Chapter 2 is to make a statement to stakeholders in order to reduce their fear or uncertainty and to avoid any indication of an unwillingness to answer questions. In Milwaukee, however, city officials waited to communicate with the public until they could attribute the sickness to a definite source.

Seeger (2006) claims that, "warnings and recalls must be issued even when some level of uncertainty exists about the exact nature of the harm. Waiting until all uncertainty is reduced usually means that the warning is simply too late" (p. 241). On April 7, after *Cryptosporidium* was found in the water, Mayor Norquist issued a boil water advisory, a "drastic public health measure" (Altman, 1993, p. 3). On April 15, after being told that tests showed the water to be safe again, he lifted the advisory, even though there were "no Federal guidelines or objective criteria for lifting an advisory against drinking untreated water after a *Cryptosporidium* outbreak" (Lawrence, 1993, p. 3). Dr. Jeffery P. Davis, the Wisconsin state epidemiologist said, "Pipes in the city water supply were flushed and samples tested. Steps were taken to reinforce good waterworks practices, and... Norquist ordered Milwaukee water quality standards tightened" (Lawrence, 1993, p. 3). After the advisory was lifted, Norquist said, "I now have renewed confidence that the drinking water in Milwaukee is safe enough to drink–safe enough to use for any purpose" ("It's water all," April 16, 1993). Beyond the statements made by Norqist, little was shared with the public until much later. Best practices in risk communication suggest that risk communicators should maintain consistent contact with the public until the risk subsides. Seeger (2006) explains, "The public has the right to know what risks it faces, and ongoing efforts should be made to inform and educate the public using science-based risk assessments" (p. 238).

On April 14, 1993, roughly four months after the initial contamination, James Kaminski, a public works official, acknowledged that "there were no policies or procedures to notify public health official about turbidity problems" ("City's water crisis," 1996, p. 4). A December 2, 1993, report put the blame on Health Department personnel:

> Thomas Schlenker, outgoing medical director of the Milwaukee Health Department, says city water department officials "passed off" early reports of water problems the previous spring and then "danced around" Health Department requests for more information. In addition, Schlenker said, chemists at the plant "were sounding the alarm bell" and telling their supervisors that something was amiss, but no one had enough experience with the new coagulant to know how to respond. (City's water crisis, 1996, p. 4)

Human errors and ineffective risk communication cost lives in Milwaukee. In an interview, Joan Rose a *Cryptosporidium* researcher, said, "I don't think anybody had a plan for *Cryptosporidium*.... It kind of got put on the back burner" ("EPA ignored warning," 1993, p. 9A). Even when officials realized what had happened they were not prepared to deal with the consequences.

The severity of the Milwaukee's Cryptosporidiosis outbreak could have been reduced if more risk monitoring systems and partnerships had been put in place. Jim Hughes, director of the CDC's National Center for Infectious Diseases, highlighted said the need for this communication: "If there had been such a system including pharmacies, nursing homes, and labs in 1993, that outbreak would have been detected much earlier than it was" (Marchione, 2003, p. 1B). Initially, however, no one knew how to respond:

On April 10, 1993, just three days after the boil water advisory was made:

> A study of AIDS patients strongly suggested five months ago that Milwaukee had a problem with *Cryptosporidium*... [b]ut health officials never found out about it, apparently because of a dispute with the local AIDS group that had the information... city health officials said angrily today that the information might have helped head off this month's epidemic. "If we had had this study earlier, it would have changed the way we look at what's in the water," said... Nannis. "It's perplexing and very frustrating that this information wasn't shared with the Health Department." The local AIDS group, the AIDS Resource Center of Wisconsin, says it did not share the information with city officials because the sample was not a systematic survey and because the organization has such a poor working relationship with health officials. (Hinds, 1993, p. 6)

As we discuss in Chapter 2, risk communicators should be coordinating and collaborating with other credible sources. Milwaukee officials were not communicating appropriately across agencies. A lack of communication added to the uncertainty surrounding the outbreak, and because of the level of uncertainty city officials failed to respond in a timely, open, and honest manner.

Crisis communicators should have "consistency of a message [because it] is one important benchmark of effective crisis communication" (Seeger, 2006, p. 240). During the Cryptosporidiosis crisis, Milwaukee city officials did not have a clear message and this left stakeholders in a greater state of uncertainty. Establishing pre-event partnerships and effective risk communication could have helped prevent many of the problems encountered in this case. What follows is a discussion of some guidelines that were established as a result of these failures.

Infuse Risk Communication into Policy Making

For the city, the Cryptosporidiosis outbreak was a learning experience. Following the outbreak, efforts were made to better prepare for the risk of future water-borne outbreaks. Such planning is essential. Providing employees with a crisis response plan can serve as a reminder of potential problems, but also provides checkpoints that may reduce the possibility of a crisis (Seeger, 2006). Such planning, therefore, can enhance overall mindfulness regarding risks. However, in order to remain viable, planning must be ongoing. Organizations should "establish strategic partnerships before a crisis occurs. These collaborative relationships allow agencies to coordinate their messages and activities" during a crisis (Seeger, 2006, p. 240). When they do, new organizational members can stay current on their roles during a crisis.

In Milwaukee, part of the planning involved coordinating relationships with agencies both large and small. For example, the CDC, U.S. Department of Health and Human Services, Milwaukee Health Department, Mayor Norquist, local pharmacies, physicians, local newspapers, and other health officials have since learned to communicate with each other. Just a few days following the Milwaukee event, Norquist pledged that Milwaukee would test regularly for the *Cryptosporidium*. He added that he has urged U.S. Health and Human services Secretary Donna Shalala to "require all water systems in the nation to do the same" ("Milwaukee seeks source," 1993, p. 1A). In 1997, Paul Nannis, Commissioner of Health for the city Health Department, reported that:

> "in reaction to the . . . outbreak, we now have the most-monitored water in the United States" . . . Nannis . . . [said] that his department has its own crypto lab . . . The health and water departments have extensive protocols for reviewing a host of parameters for water quality. . . . [and] The city also has installed state-of-the-art turbidity monitors (turbidity, or water cloudiness, often acts as a marker for water-quality problems) and particle counter at the two water treatment plants. (Wang-Cheng, 1997, p. 2)

External agencies took steps to prevent any future Cryptosporidiosis outbreaks. The Working Group on Water-borne *Cryptosporidium* at CDC published a Public Health handbook (1997) that highlights the importance of having surveillance in place:

> Local public health officials should consider developing one or more surveillance systems to establish baseline data on the occurrence of Cryptosporidiosis among residents of their community and, where possible, obtain sufficient epidemiologic data to identify potential sources of infection. . . . Surveillance should be considered by all communities whose water utility provides service to at least 100,000 persons, and whose water supply is derived from surface water. (p. 1–2)

The outbreak in Milwaukee also highlighted the need for a variety of agencies to take safety precautions. According to the U.S. Department of Health and Human Services (1995):

> [The epidemic] emphasized the need for a) improved surveillance by public health agencies to detect and prevent such outbreaks and b) coordination among interested groups and agencies to respond appropriately to such outbreaks. It also stimulated efforts to develop regulatory standards for *Cryptosporidium* in drinking water. (p. 1)

Officials started paying more attention to what needed to be done to protect the people. Epidemiology researchers Ballester and Sunyer (2000) highlighted that "two more needs have been stated as crucial for the future microbiological safety of drinking water, integration of risk assessment methodologies, and the understanding of the pathogen's ecology" (p. 4). Researchers Ballester and Sunyer (2000) also observed that more people need to be involved:

> A more proactive role of public health professionals has been demanded, encouraging the promotion of a communication programme with physicians and other sanitarian professionals, and contributing to the development of public health policies that limit contamination of source water, improve water treatment and protect public health (p. 4)

Experts have since gathered to discuss and examine health issues, monitoring of contaminants, emergency management, and communication. Many lessons have been learned:

> The cryptosporidious epidemic was a wake-up call that helped prepare local officials to detect and respond to bioterrorism if it happened... It exposed holes in the public health system-from surveillance to communication to water safety and security-that have since been plugged. (Marchione, 2001, Para. 3)

The 1993 epidemic raised water safety awareness all over the country. Milwaukee's error illustrated the importance of having open communication networks across agencies. A crisis of this nature teaches many lessons, some from the failures and some from the successes.

Implications for Effective Risk Communication

The *Cryptosporidium* case in Milwaukee was the largest water borne illness in U.S. history. Many lives were negatively affected due to Milwaukee's ineffective risk communication and pre-crisis preparedness. As a result, city and health officials have taken corrective action to ensure a future crisis situation be dealt with differently. In light of the lessons and best practices in risk and communication, this case illustrates three specific implications for effective risk communicators.

Listen for Potential Risk

Effective risk communicators are open to receiving warning signs of potential risk. In the case of Milwaukee, the city was not only ill-prepared for the crisis but also failed to account for early complaints voiced by the public. A central factor of risk communication involves effective listening. James Lee Witt, former head of the Federal Emergency Management Association, explains, "Communication, when it's working, can help you know when a crisis is coming—sometimes early enough to prevent it" (Witt & Morgan, 2002, p. 45). He argues that consumer complaints are one of the first places that risk surfaces and can be identified. Effective risk communicators consistently monitor the environment for potential risk. They take complaints seriously and pay attention to the dynamic nature of risk.

Communicate Early and Often about Risk

Effective risk communicators assess the communication context quickly and communicate openly about known hazards. Milwaukee was caught off guard after known risks emerged in their water treatment facility. They also made a key mistake in their risk communication by failing to notify the public immediately and communicate with them consistently about the harm of drinking tap water. Milwaukee's shock in learning that the safety of its water treatment process was compromised is understandable. However, it is critical for risk communicators to immediately notify the public of the known risk and maintain consistent updates with the public as long as the risk is evident. The shocking nature of risk can sometimes paralyze an organization. As a result, the perception is that the organization is unethically withholding risk information from the public. Neither perception is a model for managing risk successfully. Effective risk communicators make immediate contact with the public about risk and maintain regular contact with the public about risk levels throughout an event.

Learning is Essential to Effective Risk Communication

As a result of the *Cryptosporidium* outbreak in Milwaukee, every city in the United States changed its standards for identifying contaminated water in their treatment facilities. Effective risk communicators learn from their experience and try to enact changes in their industries. This case illustrates both experiential learning on the part of Milwaukee and vicarious learning by the CDC and the many water treatment facilities across the nation. Effective risk communication should emphasize what the organization has learned and how risk will be managed more successfully in the future.

References

Allen, S. (1995, June 24). MWRA finds microbes similar to deadly parasites. *The Boston Globe.* Metro/Region, p. 17.

Altman, L. (1993, April 20). The doctor's world: Outbreak of disease in Milwaukee undercuts confidence in water. New York Times, p. 3.

Ballester, F., & Sunyer, J. (2000, January). Drinking water and gastrointestinal disease: Need of better understanding and an improvement in public health surveillance. *Journal of Epidemiol Community Health, 54*, 3–5.

Beil, L. (1995, May 28). Tap-water bug trips alarms disease experts try to fathom parasite that defies chlorine and may be prevalent in city water supplies. *The Toronto Star*, p. F8.

City's water crisis bubbled to surface quietly, quickly. (1996, February 27). *Milwaukee Journal Sentinel*, p. 4.

Corso, P. S., Kramer, M. H., Blair, K. A., Addiss, D. G., Davis, J. P., & Haddix, A. C. (2003). Cost of illness in the 1993 Waterborne *Cryptosporidium* outbreak, Milwaukee, Wisconsin. *Emerging Infectious Diseases*. Retrieved from http://www.cdc.gov/ncidod/EID/vol9no4/02-0417.htm.

EPA ignored warnings about *Cryptosporidium*. (1993, September 20). *Capital Times*, p. 9A.

Hinds, M. D. (1993, April 10). Study hinted at a parasite problem in Milwaukee. *New York Times*, Sec. 1, p. 6.

Its water is all clear, Milwaukee is told it can stop boiling water." (1993, April 16). *New York Times*, p. 16.

Lawrence, K. A. (1993, April 20). The doctor's world; Outbreak of disease in Milwaukee undercuts confidence in water. *New York Times*, Sec. 3, p. 3.

MacKenzie, W. R., Hoxie, N. J., Proctor, M. E., Gradus, M. S., Blair, K. A., Peterson, D. E., Kazmierczak, J. J., Addiss, D. G., Fox, K. R., Rose, J. B., & Davis, J. P. (1994, July 21). A massive outbreak in Milwaukee of *Cryptosporidium* infection transmitted through the public water supply. *The New England Journal of Medicine, 331*:161–167. Retrieved September 6, 2006 from http://content.nejm.org/cgi/content/full/331/3/161.

Marchione, M. (2001, September 29). *Cryptosporidium* taught lesson on bioterrorism. Retrieved September 6, 2006, from http://www2.jsonline.com/alive/news/sep01/bio30092901.asp?format=print

Marchione, M. (2003, April 6). Lessons linger; 10 years ago, *Cryptosporidium* gripped the city. *Milwaukee Journal Sentinel*, p. 01B.

Milwaukee seeks source of bad water. (1993, April 10). St. Petersburg Times, p. 1A.

Seeger, M. W. (2006). Best practices in crisis communication: An expert panel process. *Journal of Applied Communication Research*, *34*(3), 232–244.

Ulmer, R. R., Sellnow, T. L., & Seeger, M. W. (2007). *Effective crisis communication: Moving from crisis to opportunity*. Thousand Oaks, CA: Sage.

U.S. Department of Health and Human Services, Public Health Service, Center for Disease Control and Prevention. (1995, June 16). Assessing the public health threat associated with water-borne Cryptosporidiosis: Report of a workshop. *Morbidity and Mortality Weekly Report* (*MMWR*) 1995; 44(No. RR-6): [inclusive page numbers].

Wang-Cheng, R. (1997, March 17). Important changes made or underway city water closely monitored since *Cryptosporidium* outbreak. *Milwaukee Journal Sentinel*. Health, p. 2.

Working Group on Water-borne *Cryptosporidium* (1997). *Cryptosporidium and Water: A Public Health Handbook. CDC*. Atlanta, Georgia.

Witt, J. L., & Morgan, G. (2002). *Stronger in broken places: Nine lessons for turning crisis into triumph*. New York: Times Books.

Chapter 6
Hurricane Katrina: Risk Communication in Response to a Natural Disaster

Written with Will Whiting

From wildfires in California to tornadoes in Oklahoma to hurricanes on the Gulf Coast, natural disasters are common events. When natural disasters strike, they threaten and disrupt the normal activities of businesses, churches, and schools. Hurricanes, like most natural disasters, also threaten the access that people have to basic needs like food and water. Ultimately, most natural disasters prevent rescue workers from transporting basic commodities like food and water to those in need.

Over the last several years, numerous natural disasters have threatened the United States. The greatest natural disaster of recent history that produced a significant threat to humanity was Hurricane Katrina. The impact Hurricane Katrina had on the U.S. Gulf Coast left many citizens, including children and the elderly, without access to basic necessities such as food and water. The shortage of food and water during the Hurricane Katrina disaster resulted in part from government officials failing to embrace and enact the best practices in risk communication. If a plan had been in place and the lessons of risk and crisis communication followed, the intensity of the crisis may have been abated, and communication would have enabled leaders to accurately prepare for, manage, and move past the destructive event. This case proceeds as follows: (1) an overview of the case, including a time line, is given; (2) the case is then analyzed using the best practices in risk communication; and finally, (3) practical implications for effective risk communication identified from this case are provided.

Managing Risk Communication During Hurricane Katrina

Compared with all other natural disasters ever occurring on U.S. soil, Hurricane Katrina was the most expensive. Media reports indicate that recovery costs associated with Katrina will be more than six times that of Hurricane Andrew in 1992, the storm that once held the distinction of being the most powerful hurricane to impact the continental U.S. (Associated Press, 2005b). With regard to human lives, Katrina also has the highest death toll. News clippings

T. L. Sellnow et al. *Effective Risk Communication*

DOI: 10.1007/978-0-387-79727-4_6

August 23, 2005	Tropical depression forms over the Bahamas.
August 24 – 11 a.m.	Tropical depression strengthens to form tropical storm Katrina. The storm moved across the Bahamas toward Florida. Residents of the southeastern coast of Florida are put under a tropical storm watch.
August 25 – 4 p.m.	Tropical storm Katrina officially becomes a Category 1 hurricane.
August 25 – 7 p.m.	Hurricane Katrina makes landfall in south Florida, killing nine people.
August 26 – 5 a.m.	Hurricane Katrina weakens to a tropical storm as it reenters the Gulf of Mexico and leaves Florida.
August 26 – 11 a.m.	The storm strengthens and is now classified as a Category 2 hurricane.
August 26 – 4 p.m.	National Hurricane Center (NHC) warns that Hurricane Katrina is expected to reach a dangerous Category 4 intensity before making landfall in Mississippi or Louisiana. Hours later in anticipation of the storm, Mississippi Governor Haley Barbour and Louisiana Governor Kathleen Blanco declare states of emergency.
August 27 – 5 a.m.	Hurricane Katrina reaches a Category 3 status with the Gulf Coast in its path. During the day, President Bush declares a state of emergency for Louisiana. Louisiana's residents living in low-lying areas are ordered to evacuate.
August 28 – 2 a.m.	Hurricane Katrina strengthens to a Category 4 storm.
August 28 – 7 a.m.	Hurricane Katrina strengthens to a Category 5 storm.
August 28 – 10 a.m.	As Hurricane Katrina's wind reaches 175 mph, New Orleans Major Ray Nagin orders a mandatory evacuation for everyone living in the city.
August 29 – 4 a.m.	Hurricane Katrina is downgraded to a Category 4 storm.
August 29 – 7 a.m.	Hurricane Katrina makes landfall near the mouth of the Mississippi River.
August 30	New Orleans has no power, no drinking water, dwindling food supplies, and widespread looting.
August 31	The entire region is declared a public health emergency because of contaminated, stagnant water.
September 1	Stranded residents in New Orleans remain on roofs and in the backs of trucks with no water or food. Department of Homeland Security (DHS) announces 4,200 National Guard troops will be deployed to New Orleans over the next three days.
September 2	Tired and angry stranded people welcome the first sign of food and water at the convention center in New Orleans more than four days after Hurricane Katrina made landfall. The U.S. Army Corps of Engineers (ACE) says it will take 36–80 days to completely drain the city of New Orleans.
September 3	Tens of thousands of evacuees are cleared from the New Orleans Superdome and Ernest Morial Convention Center where they had been living with very little food and water since the storm made landfall. The ACE brings in pumps and generators from around the nation to help get New Orleans' pumps back on-line.
September 4	Water and air rescue efforts continue in New Orleans. The U.S. Coast Guard said it rescued more than 17,000 people by this date, almost twice as many as it had saved in the previous 50 years combined. Helicopters dropped emergency food and water to people awaiting rescue.
September 5	Officials encourage residents remaining in New Orleans to evacuate. Deputy Policy Chief Warren Riley said that "there is no reason—no jobs, no food—no reason for them to stay." Body recovery teams conduct house-to-house searches for human remains in New Orleans, and helicopters continue search-and-rescue operations for survivors. Coast Guard reported rescuing more than 22,000 people.

Fig. 6.1 Timeline
Source: (CNN, 2007)

September 6	Public health officials report "minor outbreaks" of diarrheal diseases in children evacuated from the flood zone. The Centers for Disease Control and Prevention (CDC) said that five people died from infection with *Vibrio vulnificus*, a form of bacteria that causes cholera. An official in the New Orleans' mayor's office said that the standing floodwater is likely contaminated with *E. coli* bacteria. The U.S. House Government Reform Committee announced it will begin hearings in one week to investigate the local, state, and federal response to Hurricane Katrina and its aftermath.
September 7	Officials from the CDC and Environmental Protection Agency (EPA) encourage people to have very little contact with contaminated floodwater. Their preliminary tests showed dangerous bacteria at 10 times the acceptable level, the highest the test could measure. High levels of lead also were detected. U.S. House and Senate leaders announced the formation of a bipartisan Congressional committee to investigate the response to Hurricane Katrina at all levels of government.
September 8	In a news conference, Red Cross and Louisiana state officials said the state asked the Red Cross to delay bringing relief supplies into New Orleans.
September 9	Terry Ebbert, director of Homeland Security for New Orleans, said that early results from a sweep for bodies indicated that the number of dead people may be lower than earlier projections, which were as high as 10,000.
September 10	The mayors of Slidell, Louisiana, and Pascagoula, Mississippi, criticize the Federal Emergency Management Agency's (FEMA) assistance to their cities as slow and inadequate. Slidell Mayor Ben Morris said, "Everything we did was on our own."

Fig. 6.1 (continued)

speculated that Hurricane Andrew killed 23 people (Hurricane Katrina timeline, 1999–2000). In comparison, we may never have an accurate toll reflecting the number of deaths that resulted from Katrina, but media reports after the storm speculated that "thousands [were] feared dead" (Associated Press, 2005a). The extensive death toll that resulted from Hurricane Katrina included a diverse population, many of whom perished because they did not have food and water for several days after the storm made landfall. This fact alone helps legitimize Katrina's place at the highest position when the Federal Emergency Management Agency (FEMA, 2007).

It is difficult to imagine a storm that began as a small tropical depression evolving into a disaster of forcible magnitude, but that is what happened. On August 23, 2005, Katrina formed as a weak tropical depression near Nassau in the Bahamas. Three days later, the depression entered the Gulf of Mexico and was elevated to a tropical storm and given the name "Katrina." On that same day, weather experts at the National Oceanic & Atmospheric Administration (NOAA) issued an advisory, noting that "tracks [were] clustered between the eastern coast of Louisiana and the coast of Mississippi" thus increasing the "confidence in [the] forecast" (NOAA, 2007). In the Gulf, Katrina continued to strengthen and rarely deviated from its Northwestward track toward Louisiana and Mississippi. Forecasters continuously updated watches and warnings. The storm eventually reached Category 5 status on the Saffir-Simpson Hurricane Scale before making landfall as a powerful Category 4 storm. As the storm's power minimally deviated, the eye of the storm did not.

Katrina stayed on the track that forecasters had predicted and headed directly for the unprepared city of New Orleans and the Mississippi coast.

A short six days from its meager beginnings, on August 29, 2005, Hurricane Katrina crashed ashore near the mouth of the Mississippi River precisely where forecasters had originally predicted. Katrina destroyed cities and towns in Mississippi and Louisiana, and crippled municipalities across the Gulf Coast from Texas to Florida. Katrina devastated the infrastructure of the region and knocked out communication lines, thereby making the coordination and delivery of food and water in the disaster zone a chaotic nightmare.

As the storm pushed inland, news reports hinted that government officials and agencies had failed to coordinate resources. This left displaced citizens without food and water for several days immediately following Katrina's arrival. Officials had not planned or prepared for a crisis of Katrina's magnitude, despite predictions that forecasters made days in advance. A timeline of events in the Hurricane Katrina disaster (see Figure 6.1) clearly depicts the storm's emergence and impact.

Applying the Best Practices in Risk Communication to Hurricane Katrina

This section, first, analyzes what happened when officials and agencies neglected to extensively plan for the disaster. Second, we examine officials' poor coordination of resources and collaboration after Katrina made landfall. Third, an analysis of officials' and agencies' failure to provide honest, candid, open and accountable communication during the crisis is presented.

Infuse Risk Communication into Policy Making

Planning for a natural disaster or crisis event before it occurs is an important lesson with which all risk communicators should be familiar. Pre-planning helps offset the threat of a crisis producing significant harm. Analysis of news reports before, during, and after the Katrina natural disaster reveals that officials and agencies engaged in short-term planning processes, but they did not plan extensively for a disaster of Katrina's magnitude. Additionally, when Katrina made landfall, the short-term preparations that had been developed were not properly enacted.

The failure to engage in long-term planning and enact existing short-term plans deeply intensified the crisis. Media reports indicated that the National Weather Service (NWS) and the National Hurricane Center (NHC) precisely predicted the path of the storm and the devastation that it would bring (Associated Press, 2005d). These predictions, along with previous studies completed by experts demonstrating what would occur if a hurricane of this magnitude ever struck the New Orleans metropolitan area, provided enough time for FEMA and local and state government officials to prepare for and enact existing plans to help offset the disaster, including the National Response Plan (NRP). In addition to this problem, officials and agencies responsible for enacting the plans did not fully understand their roles and duties prior to Katrina. These problems–coupled with others–resulted in thousands of citizens stranded without food, water, and other resources necessary for survival.

National Response Plan

According to the Government Accountability Office (GAO), the NRP began circulating within government agencies nine months prior to Katrina. Its purpose was to help agencies specifically the Department of Homeland Security (DHS) and organizations that fall under the DHS umbrella–manage large scale crises like Katrina (USGAO, n.d.). The NRP is a 426-page document that incorporates lessons of crisis management and integrates them into a unified document for use in crisis management and response. The NRP features specific protocols that are to be followed in crisis situations. Some protocols related to hurricanes and natural disasters include the following:

- DHS and other government agencies "move initial response resources (critical goods typically needed in the immediate aftermath of a disaster such as food, water, emergency generators, etc.) closer" to the crisis-prone area (p. 7).
- Response actions of agencies coordinating the emergency management include providing "the provision of public health and medical services, food, ice, water, and other emergency essentials" to individuals impacted by the storm (p. 54).
- "Disaster support commodities" such as "baby food, baby formula, blankets, cots, diapers, meals ready-to-eat, plastic sheeting, tents, and water" are "pre-staged" and readily available to residents in potential catastrophic areas (p. 67).
- The Department of Health and Human Services "assists in determining the suitability for human consumption of water from local sources" after a disaster strikes (ESF #3).

These protocols, along with the numerous other policies and procedural components of the NRP, were not followed in the Katrina disaster. Emergency responders at all levels of the government failed to extensively plan for the disaster, execute existing procedural policies and ultimately mobilize food and water. For example, Gretna, Louisiana, Mayor Ronnie Harris said in an interview to New Orleans television station WWL-TV that it took FEMA "five days to get into Gretna with food and water" (Planchet, 2005). According to the NRP protocols, the commodities should have been "pre-staged" in the potential catastrophic areas. Another report from the cable television network, CNN, confirmed that the protocols identified in the NRP were not being followed in the Hurricane Katrina response. One CNN reporter in the thick of Katrina's aftermath said, "People have been sitting there without food and water waiting. They are asking... 'when are they coming to help us?' " (Lawrence, 2005, Para. 8).

Like the CNN reporter, FEMA officials recognized that their planning was poor. One senior FEMA employee working in the New Orleans Superdome, where many of the evacuees were assembling, sent an electronic mailing to his boss, then-FEMA Director Michael Brown. The mailing included the following: "...the situation is past critical...thousands gathering in the street with no food and water...estimates are many people will die within hours" (Meyers, 2005, Para. 2). This communication reinforces how poor planning intensified a risk, and pre-existing protocols in the NRP were not followed or enacted.

In addition to the discrepancies mentioned above, other reports described how FEMA took too long to move their officials to the Gulf Coast. One news report maintained that officials were "completely unprepared" making the emergency response efforts difficult (Kam & Gomez, 2005, Para. 2). Personnel were not mobilized prior to the hurricane despite the NRP protocol that implies officials will already be at the site of a potential catastrophic disaster. It took some FEMA representatives four days to arrive on the scene (Kaptur, 2005). Perhaps even more daunting is that many FEMA officials did not receive orders warranting their assistance until the day after Katrina made landfall. Even then, reports suggested that FEMA officials were given additional time to arrive despite the risk of citizens starving, dehydrating and dying. These indicators show how poor planning intensified a deadly crisis in an unprepared region.

Collaborate and Coordinate About Risk with Credible Information Sources

In addition to pre-event planning, another relevant lesson in risk communication is collaboration and coordination to assist in the establishment of partnerships. If partnerships are established in the pre-crisis phase, officials are better able to come together and manage resources when a crisis develops. This gives citizens a better chance of obtaining items they need like food, water and medical supplies. Important to note, however, is that if the planning stage is poorly conducted, the collaboration and coordination may not be successful. Looking back at the evolution of the Katrina disaster, it is now obvious that officials and agencies failed to plan and thus failed to collaborate and coordinate with each other. As a result, citizens went without food, water and other supplies needed for survival.

The lack of collaboration and coordination among officials and agencies intensified the Katrina disaster. The failure to establish partnerships meant that basic necessities like food and water that are common in the wake of any natural disaster never reached the people that were in the most critical need. An examination of a partial list of denied resources is indicative of officials and agencies failing to collaborate and coordinate with each other before, during, and after Katrina became a major disaster.

Wal-Mart Denied

Despite the looming risk, the majority of the basic living resources that people needed to survive had not been moved to the Gulf Coast prior to Katrina's landfall. Unfortunately, when these needed supplies of food and water were offered, they were often denied by officials and agencies coordinating the crisis response. Coupled with the lack of basic necessities, the insufferable heat made matters worse. For example, the media reported that Arkansas-based retail giant Wal-Mart "sent three trailer trucks loaded with water," but "FEMA officials turned them away" (Shane, 2005, Para. 14).

Truckers' Denied

Like Wal-Mart, others experienced similar frustrations. Bill Lutz, a truck driver from Wisconsin working for the U.S. Army Corps of Engineers through FEMA, had a truck loaded with water and ice destined for the Gulf Coast. According to a news report, Lutz traveled to Mississippi where he was told that the water would not be needed on the Gulf Coast and that he should make his way to South Carolina. Following those orders, Lutz drove to South Carolina only to find that there was no place to deliver the water and ice. In an interview with *The Norman Transcripts*, Lutz said, "I sound angry and I am but I hate inefficiency" (Martirano, 2005, Para. 6). This comment adequately sums up the lack of coordinated relief supplies.

Red Cross Denied

In addition to Wal-Mart's failed attempt to provide aid, other people and organizations were turned away as well. The *Pittsburgh Post-Gazette* reported that the American Red Cross attempted to deliver food to stranded individuals who had taken up residence in the New Orleans Convention Center, but Homeland Security turned them away and requested that they not return. According to American Red Cross spokeswoman Renita Hosler in a press interview, "The Homeland Security Department... requested and continues to request that the American Red Cross not come back into New Orleans" (Rodgers, 2005, Para. 3). Homeland Security must not have been aware of the repeated pleas for food and water that came in from distressed individuals who had nothing to eat or drink and as a result feared for their lives.

International Help Denied

National aid was not the only help denied; international help was denied, as well. *USA Today* reported that a German military plane carrying more than "fifteen tons of military rations" presumed to include food, water, and medical aid, was sent back to Germany after leaving for the U.S. (Associated Press, 2005c). The report indicated the plane was turned away because "it did not have required authorization" by U.S. officials to make the delivery (Associated Press, 2005c). In this instance, desperately needed resources were denied, and a door was closed on a nation reaching out to help a neighbor in dire need.

Coast Guard Denied

Food, water, and medical supplies were not the only resources denied by officials and agencies coordinating the crisis response. The media reported that a U.S. Coast Guard ship with 1,000 gallons of fuel on board was denied access to one of the docks in the devastated region. Among the reasons the fuel was shipped included helping to keep military relief trucks running and power generators in New Orleans' hospitals. According to Aaron Broussard, the President of Jefferson Parish in southeastern Louisiana, rather than accepting the fuel, FEMA officials denied it, preventing locals like Broussard from getting his share. Because FEMA did not want the Coast Guard's help, the ship with 1,000 gallons of fuel sat anchored in the ocean off of the coast for several days (Jay, 2005).

Firefighters, Police, Rescue Workers Denied

Unfortunately the list of denied resources does not end there. Emergency response teams were denied access to the region, as well. The *Chicago Tribune* reported that the city of Chicago offered to send 44 Fire Department rescue and medical personnel and their gear, more than 100 police officers, 140 Streets and Sanitation, 146 Public Health, and 8 Human Services workers, and a fleet of vehicles that included 29 trucks, two boats, and a mobile clinic. The help offered by the city of Chicago was tremendous, but FEMA officials responded with a simple message: send "only a single tank truck" (Washburn, 2005, Para. 2).

Regrettably, FEMA offered few definitive answers to explain why they turned help away. The one explanation that FEMA officials offered more than once was their fear of "security concerns" (Senate Committee, 2006). Regardless of this concern, the examples presented above signify the poor enactment of the lessons of risk and crisis communication discussed earlier in this section. Protocols found in the NRP called for the collaboration and coordination of individuals and agencies so that food and water could be pre-staged and provided to displaced citizens. Poor planning, compounded by the lack of collaboration and coordination among officials and agencies, intensified the crisis. Without basic necessities like food and water, people became ill and many died.

Present Risk Messages with Honesty

Pre-event planning and collaboration and coordination are important lessons in risk communication. Additionally, it is important for risk communicators to be honest, candid, open and accountable with their communication. These qualities establish credibility and trust in the eyes of stakeholders. An analysis of the Hurricane Katrina response revealed that leaders' actions and decisions failed to demonstrate these qualities.

Those managing the response in the aftermath of Hurricane Katrina neglected to provide honest accounts of their behavior. Officials could not deny that after Katrina food and water resources were unavailable. Media reports from the storm-ravaged region portrayed stranded citizens starving and pleading for food and water. *USA Today* described Fox News reporter Shepherd Smith's report of irresponsibility demonstrated by government officials. Government officials said to storm victims, "You go here, and you'll get help," or "You go to the Superdome and you'll get help" (Johnson, 2005, Para. 2). But, Smith said, in reality, "They didn't get help. They got locked in there. And they watched people being killed around them. And they watched people starving. And they watched elderly people not getting any medicine" (Johnson, 2005, Para. 2).

The response to Hurricane Katrina was problematic and contrary to sound lessons in risk communication. Rather than accounting for the shortages and honestly explaining why food and water had not been pre-staged prior to the storm, officials offered excuses and shifted blame. In a press briefing, White House press secretary Scott McClellan defended the Administration; he acknowledged, however, that officials at all levels of the government had participated in "the blame game" (McClellan, 2005, Para. 23). Louisiana Governor Kathleen Blanco blamed New Orleans Mayor Ray Nagin for the lack of food, water, and other resources. Nagin blamed FEMA. Despite the obvious breakdown in responsibilities and communication, government officials and agencies on the ground and in Washington, D.C., were rarely willing to admit that failure occurred. New York University journalism professor, Jay Rosen, believes that the media covering Katrina seemed "to be much more effective than the Administration in representing the public's reaction to the disaster" (Johnson, 2005, Para. 10). Had there been honest, candid, and open communication from the officials and agencies managing the Katrina response, perhaps Rosen's and similar claims would never have been leveled.

It is equally important to point out that the "blame game" also resulted in some officials losing control of their own emotions in the midst of the chaos. *Time Magazine* photographed Governor Blanco sobbing helplessly. Ultimately, this signified to stakeholders that officials were in a state of distress, with officials from various agencies engaged in blaming each other while trying to make sense of the crisis management efforts (Tumulty, 2005, p. 38). Blanco, however, was not the only person to lose her composure. During an interview on national television, Nagin denied responsibility for his city's problems. Nagin said the federal government needed to "get off [their] asses and . . . do something" (Robinette, 2005).

Tardy and insufficient information resulted in many unanswered questions. Ultimately, the failure to communicate in an honest, candid, open, and accountable manner damaged the credibility of the federal government and everyone who attempted to manage the situation. The resulting problems caused citizens to question the ability of officials and agencies to manage the recovery efforts.

Implications for Effective Risk Communication

Poor risk planning and communication response to Hurricane Katrina will be discussed for decades. There should be no doubt that the Katrina disaster intensified in part because officials and agencies failed to embrace the risk and communication principles discussed in this case. The previous analysis of officials' use of communication strategies provides several important implications for all risk communicators. Understanding and embracing the following implications will enhance our ability to plan for and respond to natural disasters.

Infuse Risk Communication into Policy Making

First, pre-event risk plans must be comprehensive but easily enacted. Plans should be discussed exhaustively by all individuals in the organization, especially those in key leadership roles. Leaders with expert knowledge in specific areas should be included in the planning process. For example, because natural disasters almost always involve issues related to food and water, representatives from the food and water industries should be included in the planning process. Sadly, the Katrina disaster provides an excellent example of what happens when such plans are not in place. Consider, for example, the National Response Plan (NRP). Despite NRP's existence prior to Katrina, the 426-page document was too large for officials to completely understand and implement. Officials' inability to understand each policy and the policies' requirements only added confusion to an already chaotic situation. One specific NRP policy calls for "disaster support commodities" such as "baby food, baby formula, blankets, cots, diapers, meals ready-to-eat, plastic sheeting, tents, and water" to be "pre-staged" and readily available to residents in potential catastrophic areas (USGAO, n.d.). This policy is warranted and makes sense, but many processes such as this one were overlooked due to the length of the entire document. This oversight resulted in failure for those managing the Katrina disaster. While it is not appropriate to indicate a specific length for any risk planning document, it is recommended that organizations create and adopt plans covering all aspects of a possible crisis. Most importantly, organizations should develop plans that are comprehensive of their operations and needs, and practice those plans often by simulating mock crises so that everyone in the organization is familiar with the processes involved.

Meet Risk Perception Needs by Remaining Open and Accessible to the Public

To maintain and reestablish relationships with the public, risk communicators must be honest and open with stakeholders when disseminating information. It is important for honest and open communication to exist before, during, and after a crisis. Seeger et al. (2003) argue that "communication in the form of dialogue is essential for organizations to recondition" their response efforts (p. 79). Open communication allows citizens to better understand the conditions of the situation. The demonstration of the principles of honesty and openness when disseminating information reaffirms a risk communicator's credibility and legitimacy.

In the Katrina disaster, this lesson was rarely followed when officials and agencies disseminated information. Media outlets, not government officials, were the first to report the Gulf Coast's need for food and water. Moreover, officials immediately engaged in a process of blaming each other rather than providing explanations of the situation. This adversely affected the response managers' credibility. Had they retained their credibility, congressional hearings on the Katrina disaster response might have been avoided. Without honest, candid, open and accountable communication, risk intensifies and may lead to a crisis.

Collaborate and Coordinate About Risk with Credible Information Sources

Risk managers must focus on building strong relationships with public and private stakeholders. Establishing partnerships and relationships with stakeholders reduces the risk of developing an organizational structure that constrains crisis management. Stakeholders may have ideas, resources, or services that can help responders. For example, failure to collaborate and coordinate in the Katrina response resulted in the denial of many desperately needed resources such as water, meals, ice, and foreign aid. One thousand gallons of fuel that could have powered emergency vehicles never left the ship docked off of the Louisiana coast. Medical and rescue personnel offered by the city of Chicago were turned away. When a crisis with the complexities of Hurricane Katrina occurs and people are dying because they are starving or dehydrating, turning away aid for people in dire need is unacceptable. As crisis managers, it was officials' and agencies' job to provide aid to those in need. This did not happen, however, because relationships with multiple stakeholders did not exist prior to the storm.

Effective risk communicators understand the value in partnering with other industries, establishing solid relationships with community leaders and organizations who may be able to aid in crisis management, and then communicating often with identified stakeholders. Risk planners must identify the resources that will be needed in a crisis, and then invite industries to help plan for and create manageable responses. Risk communicators should realize that industries may be more effective in providing resources crises demand. As industries are invited to provide solid commitments in the partnership, a database of suppliers should be established and made accessible from many different locations. In the Katrina disaster, relationships with community leaders and organizations were lacking and a database of resource suppliers was nonexistent. Prior to Katrina, the City of New Orleans could have partnered with water bottling plants to expedite the delivery of much needed water to the area. The city could have partnered with public schools for the use of school buses to evacuate refugees. These examples are indicative of the successful crisis management that can occur if partnerships exist with community-linked industries. Partnerships maximize resources, and allow community stakeholders to become active agents of change at a time when change is needed the most.

Failure to engage in the best practices in risk communication can be detrimental. Failure to plan for collaboration and coordination during a crisis can increase recovery effort costs. Within one year after the hurricane, the Katrina recovery costs had already exceeded $200 billion. Economic analysts predicted that the number would reach at least $300 billion (Associated Press, 2005b). While those large numbers have a definite impact on the federal budget, they pale in comparison with the costs in terms of human lives lost. The death toll from Hurricane Katrina was extremely high, and an accurate number of deaths from the storm may never be known. Sadly, many of the deaths were the unfortunate result of poor planning and management by officials and agencies. While monetary debt can be recovered, human lives lost in this storm and its aftermath will never be regained.

References

Associated Press. (2005a, September 3). Official death toll to take a long time. *The Seattle Times*. Retrieved January 31, 2007, from http://seattletimes.nwsource.com/html/hurricanekatrina/2002468585_katdeaths03.html.

Associated Press. (2005b, September 10). Katrina may cost as much as four years of war. MSNBC. Retrieved January 31, 2007, from http://www.msnbc.msn.com/id/9281409/.

Associated Press. (2005c, September 10). German plane with Katrina food aid turned away. *USA Today*. Retrieved January 31, 2007, from http://www.usatoday.com/news/nation/2005-09-10-german-aid_x.htm.

Associated Press. (2005d, September 19). Katrina's forecasters were remarkably accurate. MSNBC. Retrieved January 31, 2007, from http://www.msnbc.msn.com/id/9369041/.

CNN. (2007, March 25). Hurricane Katrina Timeline. Retrieved March 25, 2007, from http://www.cnn.com/SPECIALS/2005/katrina/interactive/timeline.katrina.large/frameset.exclude.html.

Federal Emergency Management Agency. (2007, January 29). Declared disasters by year or state. Retrieved January 31, 2007, from http://www.fema.gov/news/disaster_totals_annual.fema.

Jay, D. O. (2005, September 6). FEMA turned away aid, rescue crews, cut emergency communication lines: Witnesses. *The Dominion Paper*. Retrieved January 31, 2007, from http://dominionpaper.ca/international_news/2005/09/06/fema_turne.html.

Johnson, P. (2005, September 5). Katrina rekindles adversarial media. *USA Today*. Retrieved February 15, 2007, from http://www.usatoday.com/life/columnist/mediamix/2005-09-05-media-mix_x.htm.

Kam, D., & Gomez, A. (2005, September 10). Lack of plan hurt Katrina–hit states' response. *Palm Beach Post-Gazette*. Retrieved January 31, 2007, from http://www.palmbeachpost.com/storm/content/state/epaper/2005/09/10/m1a_response_0910.html.

Kaptur, M. (2005, September 7). Office of U.S. Congresswoman Marcy Kaptur. FEMA's ineptitude in the aftermath of Hurricane Katrina. Retrieved January 31, 2007, from http://www.kaptur.house.gov/Speech.aspx?NewsID=1435.

Lawrence, C. (2005, September 01). Stories of heartbreak and hope in Katrina's wake: Living like animals. *CNN*. Retrieved February 7, 2007, from http://www.cnn.com/2005/WEATHER/09/01/scene.blog/index.html.

Martirano, M. D. (2005, September 17). Truckers say hurricane relief shipments turned away. *The Norman Transcripts*. Retrieved February 15, 2007, from http://www.normantranscript.com/localnews/local_story_260011812/resources_printstory.

McClellan, S. (2005, September 6). The White House. Press briefing by Scott McClellan. Retrieved February 7, 2007, from http://www.whitehouse.gov/news/releases/2005/09/20050906-5.html.

Meyers, L. (2005, October 19). FEMA e-mails document disconnect on Katrina: Was former director Brown unaware how bad conditions really were? MSNBC. Retrieved February 7, 2007, from http://www.msnbc.msn.com/id/9756145/.

National Oceaning and Atmospheric Administration. (2005, August 23). National Oceanic and Atmospheric Administration (NOAA) National Hurricane Center Hurricane Katrina Forecast Timeline. Retrieved January 31, 2007, from http://commerce.senate.gov/pdf/Katrina_NOAA_Timeline.pdf.

Planchet, T. (2005, September 6). Updates as they come in on Katrina. *WWL-TV*. Retrieved February 7, 2007, from http://www.wwltv.com/local/stories/WWLBLOG.ac3fcea.html.

Robinette, G. (2005, September 2). Mayor to feds: 'Get off your asses.' CNN and WWL-AM. Retrieved October 10, 2006, from http://www.cnn.com/2005/US/09/02/nagin.transcript.

Rodgers, A. (2005, September 03). Homeland Security won't let Red Cross deliver food. *Pittsburgh Post-Gazette*. Retrieved January 31, 2007, from http://www.post-gazette.com/pg/05246/565143.stm.

Seeger, M. W., Sellnow, T. L., & Ulmer, R. R. (2003). *Communication and organizational crisis*. Westport, Connecticut: Praeger Publishers.

Senate Committee on Homeland Security and Governmental Affairs (2006, May). Hurricane Katrina: A nation still unprepared (Executive Summary, pp. 1–21). Retrieved January 31, 2007, from http://hsgac.senate.gov/_files/Katrina/ExecSum.pdf.

Shane, S. (2005, September 5). After failures, government officials play blame game. *New York Times*. Retrieved January 31, 2007, from http://www.nytimes.com/2005/09/05/national/nationalspecial/05blame.html?ex=1283572800& en=1d14ebfbd942a7d0& ei=5090.

Tumulty, K. (2005, September 19). The Governor: Did Kathleen Babineaux Blanco make every effort to get federal help? *Time Magazine, 166*, 38–39.

United States Government Accountability Office. (n.d.). National Response Plan. Retrieved January 31, 2007, from http://www.gao.gov/htext/d06808t.html.

Washburn, G. (2005, September 2). Daley 'shocked' a federal snub of offers to help. *Chicago Tribune*. Retrieved January 31, 2007, from http://www.chicagotribune.com/news/local/chi-050902daley,1,2011979.story.

Chapter 7
New Zealand Beef Industry: Risk Communication in Response to a Terrorist Hoax

Written with Kathleen Vidoloff

Hoaxes and terrorist threats, while not actual crises, can cause alarm and anxiety. Hoax terrorist attacks can be extremely disruptive; this fact is recognized by the United States Congress. In 2001, the United States House of Representatives created the Anti-Hoax Terrorism Act (2003), legislation that "makes it a felony to perpetrate a hoax related to biological, chemical, nuclear, and weapons of mass destruction attacks" (p. 3). A response to a hoax can, however, provide an experiential learning opportunity for an organization. In a hoax response, organizations respond as if a real threat is present, testing the organization's risk and crisis preparedness.

This case analyzes how the New Zealand Ministry of Agriculture and Forestry (MAF) managed risk during a hoax in May 2005. This case examines the organization's successful application of key best practices in risk communication. In addition, this case illustrates how organizations can vicariously learn from a high risk situation. This case proceeds as follows: (1) an overview of the case, including a time line, is given; (2) the case is then analyzed using the best practices in risk communication; and finally (3) practical implications for effective risk communication identified from this case are provided.

Managing Risk Communication During a Terrorist Hoax in New Zealand

On May 10, 2005, New Zealand's prime minister received a letter claiming a deliberate release of the foot and mouth disease virus on Waiheke Island. New Zealand's MAF considered the threat to be viable and responded quickly to disseminate information to media outlets regarding farmers who could have been affected by the virus ("Biosecurity threat," 2005). MAF created a website, held daily press conferences, and involved many industry organizations in the emergency response effort to address, specifically, farmers' concerns, but also general public concerns. Eventually, MAF deemed the situation a hoax, thus scaling back the crisis response, and life returned to normal for people living on Waiheke Island. As a hoax situation,

T. L. Sellnow et al. *Effective Risk Communication*

DOI: 10.1007/978-0-387-79727-4_7

this scenario (see Figure 7.1) provides the New Zealand government and other countries with agriculturally-based economies with a unique opportunity to learn more about the communication response strategies that were used. Fortunately, New Zealand has never had a case of foot and mouth (Biosecurity New Zealand, n.d.).

May 10	MAF receives letter stating that vial of foot and mouth disease virus has been released on Waiheke Island. Direct location not known. MAF releases press release on threat investigation and provides question and answer document on situation. 18 of 39 farmers on Waiheke Island are contacted.
May 11	MAF holds media conference to inform public of its efforts. 30 of 39 farmers on Waiheke Island are contacted. Hotline established for farmers to call if not yet contacted.
May 12	MAF holds media conference stating that no symptoms of foot and mouth have been found.
May 13	MAF holds media conference. Press calls today "D-Day" as symptoms would begin to show if foot and mouth disease has been released on the island. MAF launches awareness campaign on foot and mouth disease symptoms.
May 14	No foot and mouth disease symptoms appear.
May 15	No foot and mouth disease symptoms appear.
May 16	Local newspaper receives second hoax letter. MAF's press release officially declares situation a hoax. Average foot and mouth disease incubation period passes.
May 17–24	MAF does not issue any press releases. MAF reduces crisis response teams.
May 25	MAF's *Operation Waiheke* officially stands down.

Fig. 7.1 Timeline

Applying the Best Practices in Risk Communication to the Terrorist Hoax in New Zealand

In the case of New Zealand's *Operation Waiheke* response, the most salient lessons are managing crisis uncertainty, communicating with stakeholders, crisis leadership and organizational learning. These key lessons stand out during MAF's response as they were overtly demonstrated during the hoax response.

Account for the Uncertainty Inherent in Risk Communication

A key lesson in risk and crisis communication is training and preparing the organization to deal with uncertainty, threat, and communication demands before the crisis occurs (Ulmer et al., 2007). Even if an organization has a crisis plan, that specific plan may not provide an appropriate response. The organization simply cannot know everything. The situation on Waiheke Island created a high amount of uncertainty for MAF, the Federated Farmers, trading partners, farmers, and many others. To reduce the uncertainty of the hoax, MAF partnered with the New Zealand (NZ) police to determine the validity of the threat ("Disease alert," 2005; "Disease checks," 2005; "NZ police," 2005; "Quick action," 2005; Sell & Berry, 2005). The NZ police contacted additional international policing agencies, including the United States Central Intelligence Agency, to verify the legitimacy of the letter. "They came back with a high probability ranging of almost to 90–100% it was a hoax" (Biosecurity Director, personal communication, April 12, 2006).

MAF's Biosecurity Director's message, "We believe this to be a hoax, but the claim is being taken seriously," was continually reported by the media throughout *Operation Waiheke* ("Action," 2005; "All signs," 2005; "D-day for foot," 2005; "Disease checks," 2005; "Foot and mouth alert," 2005; "Foot and mouth would," 2005; Kay, 2005b; "Kiwi shakes off," 2005; "Letter sparks," 2005; "NZ foot and mouth scare," 2005; "Quick action," 2005; "Roundup: Foot and mouth scare," 2005; "Scare," 2005; "Stockfeed appeal," 2005; Waiheke farmers," 2005).

Although MAF does have risk management and crisis response plans, officials never planned or trained for a hoax response. "We had never [sic] before had ever really even considered the idea of a hoax. I guess we didn't know for sure this was a hoax, it was a serious threat" (Post-clearance director, personal communication, March 22, 2006).

After the receipt of the second letter, the police deemed the situation a hoax. Due to the 14-day incubation period for foot and mouth disease, MAF continued checking for disease symptoms ("Action," 2005; Kay, 2005a; "NZ foot and mouth scare," 2005; "Quick action," 2005; "Roundup: Foot and mouth scare," 2005; "Second letters," 2005; "Stockfeed appeal," 2005). MAF responded to the hoax as if it was a real threat, but many staff members did not think the threat was valid: "We were acting as if it was real, but understanding the chances were quite high it was a hoax" (NZFSA Communication Director, personal communication, March 21, 2006).

The repetition of MAF's hoax message and the lack of foot and mouth disease symptoms reduced the uncertainty for those on Waiheke. As a part of the response, MAF veterinarians conducted various tests on the animals looking for any signs or symptoms of foot and mouth disease. A wife of a Waiheke farmer told a local network media outlet that the local community believes the situation was a hoax: "Yes, because so far nothing has come back" ("Waiheke farmers," 2005).

Uncertainty and ambiguity were prevalent during the Waiheke response. The police investigation, the repetition of the hoax message, and the lack of foot and mouth symptoms helped reduce the uncertainty of the situation. The receipt of the second letter confirmed the hoax and reduced all uncertainty.

Infuse Risk Communication into Policy Making

For MAF, *Operation Waiheke* confirmed that response policies and procedures cannot be prepared easily or quickly in the middle of an emergency (Biosecurity

New Zealand, n.d.) In the months before *Operation Waiheke*, MAF conducted Exercise Taurus, a training exercise (Hotere, 2005; New Zealand Ministry of Agriculture and Forestry, 2006; Post-clearance Director, personal communication, March 22, 2006). During this simulation, the government's response plan was activated. Other government agencies work with MAF to determine the response procedures to the hoax. This interagency coordination creates "a whole of government response" by allowing multiple agencies to meet and decide the best way to respond to a situation (Post-clearance director, personal communication, March 22, 2006).

According to MAF's Post-clearance director, MAF's crisis response plan is broken into two parts. "One part is foot and mouth disease specific policies and procedures and the second part is just additional response [sic] procedures which is really generic procedures for how we deal with any emergency" (New Zealand Ministry of Agriculture and Forestry, 2006; Post-clearance director, personal communication, March 22, 2006). Even though the government simulated a crisis during Exercise Taurus, New Zealand Food Safety Authority (NZFSA) did not test their communication networks (NZFSA Communications Director, personal communication, March 21, 2006). MAF's Post-clearance director described the importance of training simulations: "One of the things that helped us get those plans together was the exercises that we do, you know, the simulation. And see, we had a simulation just a month before the hoax occurred, so we'd done a lot of work trying to get those plans in place to be ready for those exercises" (personal communication, March 22, 2006).

Collaborate and Coordinate About Risk Communication with Credible Information Sources

In Chapter 2 we argue that organizations should work before a crisis to cultivate strong partnerships with stakeholders. This lesson suggests creating relationships with organizations, such as media outlets or key stakeholders, which could distribute information to the public and other stakeholders during the crisis.

The partnerships among MAF, NZFSA, and the NZ police were highly valuable in the hoax response due to high economic risk to the country and the uncertainty of the situation. The agencies worked together to disseminate information and reduce the ambiguity of the hoax. This working partnership is demonstrated through press conferences, MAF media releases, and media coverage ("Biosecurity threat," 2005; "Current Waiheke," 2005; "Disease alert," 2005; Kay, 2005a; "Letter sparks," 2005; "Media advisory," 2005a; 2005b; "*Operation Waiheke*," 2005a; 2005c; Roundup: Foot and mouth scare, 2005; "Scare," 2005; "Second letters," 2005). The Biosecurity Director, the NZFSA Director and the Police commissioner all attended the live press conferences. MAF's Biosecurity Director added, "In fact, I'm confident that we were actually working together here. We weren't working in isolation, but all of our press conferences we fronted up the three of us and we each made a statement," (Personal communication, April 12, 2006).

The partnership between MAF and the locals on Waiheke was also important to the hoax response. MAF's crisis plan calls for a local community liaison group to be established in the event of a crisis or emergency. However, setting up the group takes about six weeks. In order to adapt to the situation, MAF partnered with local community chairs to support the ground response (Director Post-clearance, personal communication, March 22, 2006; "Life on Waiheke," 2005; MAF Communications Director, personal communication, March 30, 2006). The Biosecurity director stressed the importance of working with local people: "The islanders on Waiheke Island are very proud, very strong people and without their sole support, I think we would have gotten no real sense of control" (personal communication, April 12, 2006). The head of local response on Waiheke further enforced the idea of "partnership in practice" and that the team effort between MAF and Waiheke islanders was essential in keeping foot and mouth off of the island (Community chair, personal communication, May 6, 2006).

The relationship with the media played a key role in the response. In New Zealand, the relationship between the media and the government is comparatively different than in the United States. In New Zealand there is press freedom mechanism which allows the government to tell the media specific items of information to withhold from the public.

> In the case of Waiheke Island hoax, there were some things in the initial hoax letter that the police wanted to withhold because they believed that the public could comprise their ability to investigate. We had that conversation with the media and they agreed, they agree to go along with that. (MAF Communications Director, personal communication, March 30, 2006)

Due to this press procedure, MAF was able to show the hoax letter to the media and ask them not to publish it; in turn, the media agreed with the government (MAF Communications Director, personal communication, March 30, 2006; Post-clearance director, personal communication, March 22, 2006).

MAF used the media to primarily disseminate information about the current situation and the ongoing response effects. Through the media, New Zealand farmers, industry groups and trading partners learned about the hoax at a live press conference ("Farmers left," 2005; "Ferries," 2005; Office of MAF Director Post-clearance, 2006; 'The temperature," 2005). MAF did not alert farmers or trading partners before the media conferences for two reasons. First, the fear of insider trading by trading partners. Second, locating all of the farmers individually is difficult (Director Post-clearance, personal communication, March 22, 2006; MAF Communications Director, personal communication, March 30, 2006).

Relationships with the trading partners and members of the Federated Farmers organization aided the response. Both of these stakeholders disseminated information via organizational websites and emails (MAF Biosecurity Director, personal communication, April 12, 2006; MAF Communications Director, personal communication, March 30, 2006; NZFSA Communications Director, personal communication, March 21, 2006; Post-clearance director, personal communication, March 22, 2006). The Federated Farmers also organized feed supplies for farmers on Waiheke ("Disease checks," 2005; "Foot and mouth alert," 2005; "Stockfeed appeal," 2005; "Quick action," 2005). "It was actually sort of all of New Zealand pulling together against this major risk to the country," (Biosecurity Director, personal communication, April 12, 2006).

Forming partnerships with NZFSA, the NZ police, Waiheke islanders, and the media enabled MAF to communicate the hoax message to a variety of different audiences. Providing information to multiple stakeholders reduced the uncertainty of the event and created a trusting environment. The unique relationship between the NZ government and the media also played a role in the hoax response.

Involve the Public in the Dialogue About Risk Communication

The general public will have legitimate concerns about the ongoing risk and may be able to provide information to the organization. MAF's partnership with the people on Waiheke provided the government with an opportunity to have a two way dialogue during the hoax response. The two-way dialogue did not begin until later in the response.

On the first day of the response, MAF told all of New Zealand about the hoax via the media. Initially, farmers were upset as they had not been contacted by MAF before the media announcement. For example, one news article reported a Waiheke farmer saying, "I've had a few calls from journalists, but haven't heard from MAF" ("Farmers left," 2005). Another media report added that "some farmers only found out about the possible release from the media" ("Ferries," 2005). It was important for MAF to include the islanders in the response, as MAF Biosecurity director said previously, "The islanders on Waiheke Island are very proud, very strong people and without their sole support, I think we would have gotten no real sense of control" (personal communication, April 12, 2006).

Through MAF and the local community chair, public meetings were organized and held on the island ("Life on Waiheke," 2005; "Operation Waiheke," 2005b; 2005c; "Quick action," 2005; "The temperature," 2005). These meetings provided an opportunity for MAF officials to discuss the actions they were taking and the rationale for them. The meetings also provided a platform for individuals on the island to voice their concerns on the response and the situation. Several concerns included the lack of feed on the island, the controlled notice on animal movements, and their late notification of the situation ("Farmers left," 2005; "Ferries," 2005; "The temperature," 2005). Due to some of the concerns voiced, MAF sent additional people to the island (Post-clearance director, personal communication, March 22, 2006). In addition to the public meetings, the Biosecurity Director and the Minister of Agriculture went out to the island and personally met with the farmers (Biosecurity Director, personal communication, April 12, 2006; Office of MAF Director Post-clearance, 2005).

Even though MAF's response did not begin as a two-way flow of information, MAF was able to adapt their risk communication and created a dialogue between the government and the people on Waiheke. Public meetings also allowed the public to voice their concerns about the response. In turn, the government was able to make minor changes to their current response, including the assignment of additional staff to the island.

Meet the Risk Perception Needs by Remaining Open and Accessible to the Public

Being accessible to the media includes designating a spokesperson to provided information to the media, updating website posts, or sending email alerts. Refusing to comment is not a preferred action in risk communication. The key spokespeople in *Operation Waiheke* (the Biosecurity Director, the Director of NZFSA and the Police commissioner) were also the key decision makers in the response. In order to reduce the number of press calls and enable the spokespeople to make response decisions, MAF and NZFSA worked with the media's deadlines when scheduling press conferences (Office of MAF Director Post-clearance, 2005; Sell & Berry, 2005). MAF's Communications Director said, "We were running press conferences 1 hour prior to the major news deadlines, and the period of the day in the early stages and then wound back as the emergency started to wind down to one a day" (Personal communication, March 30, 2006).

MAF posted media advisories on its website alerting journalists of upcoming press conferences ("Controlled area," 2005; "Media advisory," 2005a, 2005b; "*Operation Waiheke*," 2005c; "Second letters," 2005). The media releases also included information on the status of the response and any additional updates.

The local community chair also remained accessible to the media although he was not included in the press conferences ("Life on Waiheke," 2005; "*Operation Waiheke*," 2005c). "I was fair game for the media, but I always felt confident that I was well able to handle the most difficult questions due to the avenues of communication open to me" (Personal communication, May 6, 2006).

Being accessible to the media not only includes representatives of the responding organization, but also any spokespeople involved in the response. MAF was able to meet the needs of the media by scheduling press conferences around media deadlines and continually updating their website information. By doing so, MAF partnered with the media to disseminate information about the situation.

Present Risk Messages with Honesty

The organization needs to be truthful about the ongoing crisis and not hide or refuse to comment. For MAF, "the response to the Waiheke hoax demonstrated how important it was to 'front foot' the issue" and provide frequent factual updates on the response to media, industry, and trading partners (Biosecurity New Zealand, n.d.). Communicating a hoax message proved to be a difficult task for MAF officials (Biosecurity director, personal communication, April 12, 2006; MAF Communications Director, personal communication, March 30, 2006; "Vets, police," 2005). To lessen the difficulty of delivering the hoax message, open and honest communication created a trusting environment between the government and all of New Zealand. "We made a very conscious decision early on that we would share everything we possibly could so you know any major decisions that we made we tried to share that, not just what the decision was, but the rationale for it" (Post-Clearance director, personal communication, March 22, 2006). To counteract any potential negative consequences of a hoax, MAF officials communicated openly and honestly: "I think the genuineness of the way that we communicated, the openness and the transparency, we did get the confidence of the stakeholders" (Biosecurity Director, personal communication, April 12, 2006).

MAF officials believed past behavior partnered with open and honest communication would enforce trust in the reputation of the country and in turn convince trading partners to continue working with New Zealand (Director Post-clearance, personal communication, March 22, 2006). The trading partners had to trust MAF's claim that the event was a hoax or the New Zealand dollar would drop and the economy would be affected (Director post-clearance, personal communication, March 22, 2006; NZFSA Communications Director, March 21, 2006). MAF was concerned that the hoax would not only hurt the country's economic foundation, but also negatively affect New Zealand's reputation: "We're dealing with, ultimately, the reputation of New Zealand. You know, reputation is something that you just don't mess with" (MAF Communications Director, personal communication, March 30, 2006).

MAF, NZFSA, and Federated Farmers understood the value of transparent messages, but one example fully demonstrates this principle:

> [A]s soon as you start differentiating messages you start having problems because you, it's really hard to know what you told to different people. And also different stakeholders know the same people and you know so you've got to be very careful you don't get any dissonance in the in the message. (MAF Communications Director, personal communication, March 30, 2006)

MAF made a conscious decision to incorporate the principles of open and honest communication into the hoax response. Open and honest communication creates a trusting environment for those involved in the response and has the ability to strengthen the reputation of the organization handling the response. MAF and NZFSA provided many examples on how open and honest communication was used during *Operation Waiheke*.

Treat Risk Communication as a Process

Intense risk creates stress and uncertainty. Thus, the organization should communicate an empathetic attitude to stakeholders during this delicate situation. By listening to the public concerns, the organization can address those concerns in the strategic response through communicating compassion. Communicating compassion and empathy in a hoax is quite different than in an actual crisis. MAF's spokespeople had received risk communication briefings on how to respond during a crisis (MAF Communications Director, personal communication, March 30, 2006; NZFSA Communications Director, personal communication, March 21, 2006).

MAF officials expressed concern and appreciation for the patience of islanders on Waiheke (Biosecurity Director, personal communication, April 12, 2006; Post-clearance director, personal communication, March 22, 2006). NZFSA communicated compassion especially when dealing with an alarming situation such as the risk of a foot and mouth outbreak. "We do recognize that when you communicate on the subject that people are frightened, and that you do need acknowledge their fears" (NZFSA Communications Director, personal communication, March 21, 2006).

After *Operation Waiheke*, the local community chair expressed gratitude for MAF's empathic messages: "Your sensitivity in all aspects of this engagement was respected by everyone especially our farmers. When final clearance was given and life could return to normal, we all felt that you had become part of the island family" (personal communication, May 6, 2006).

It is important to note that communicating compassion in a hoax will be different from communicating compassion in an actual crisis. Spokespeople communicated empathic and reassuring messages during *Operation Waiheke*. In turn, the Waiheke community chair expressed his appreciation for MAF's empathic messages.

Acknowledge Diverse Levels of Risk Tolerance

This lesson encourages organizations to provide self-efficacy statements, which gives the public something meaningful to do. Self-efficacy can be as simple as calling the local police department with information regarding the crisis. MAF provided many opportunities for Waiheke farmers and the public to engage in self-efficacy. MAF's media advisories, public awareness campaign, and other news articles include a hotline number for individuals to call with questions or to report foot and mouth symptoms ("Action," 2005; "Current Waiheke," 2005; "MAF braces," 2005; "Media advisory," 2005b; Office of MAF Director Post-clearance, 2005; "*Operation Waiheke*," 2005b; 2005c; "Quick action," 2005).

MAF sent staff to visit the sales yards and talk to farmers about the situation. The staff handed out leaflets and foot and mouth disease sheets ("D-day for foot," 2005; MAF Post-clearance director, personal communication, March 22, 2006). People on commuter ferries between Auckland and the island could listen to a "sound track" updating them on the hoax response (MAF Communications Director, personal communication, March 30, 2006). Additionally, individuals were invited to attend the public meetings on Waiheke Island ("Life on Waiheke," 2005; "Quick action," 2005; "The temperature," 2005).

MAF provided many opportunities for people to engage in meaningful activities. People could listen to information while commuting to work, attend a public meeting, or read a foot and mouth disease fact sheet. MAF even launched a national awareness campaign educating people on the symptoms of foot and mouth disease, which enabled the public to look for, identify, and report any foot and mouth symptoms.

Learning through a Hoax

While risk management and crisis response plans are essential, if they are outdated they are inadequate. After Exercise Taurus, the NZFSA was in the process of making a few changes to their crisis management plan when the hoax letter was received (Personal communication, March 21, 2006). MAF's Director Post-clearance admitted that his organization may not continually update the plan: "We probably don't update it formally as regularly as we should," but he said their response plan would undergo a complete review to incorporate any lessons learned from Waiheke (New Zealand Ministry of Agriculture and Forestry, 2006; Post-clearance Director, personal communication, March 22, 2006). The communication plan review focused on the following areas: communication tools, stakeholder relationships, media relationships, management systems, and management resources (MAF Communications Director, personal communication, March 30, 2006). When updating the crisis response plan, the Biosecurity director advocates for less dependence on key people as communicators and decision makers–individuals should fill one role or the other; not both (Biosecurity director, personal communication, April 12, 2006).

This lesson is crucial for organization learning. MAF will be able to look back and see what procedures worked in a hoax response and which did not. Since no actual crisis occurred, this response provided a unique opportunity for MAF to test all of their response operations. Additionally, other organizations will be able to vicariously learn from MAF's response by learning which strategies were utilized by the organizations and how those may have enhanced or not enhanced the response.

MAF illustrated key lessons in risk and crisis communication, and this case provides an opportunity for other organizations to learn vicariously from this crisis response. This section provides suggestions for organizations when training and preparing for a crisis response.

A hoax can create a high level of uncertainty. An ambiguous hoax coupled with the fourteen-day incubation period for foot and mouth disease created a high stress environment for the people of New Zealand. Partnering with the NZ police and other international agencies to investigate the hoax letter aided in the uncertainty reduction. The lack of symptoms and the receipt of the second letter also reduced the uncertainty of the hoax. It is important for agencies to understand that uncertainty plays a large role in hoax events.

Remaining open and honest about known information builds credibility for the organization. Open and honest communication by MAF officials added to the trusting environment. Although the agency withheld the hoax letter from the public, their hoax response and rationale for their actions bolstered their credibility. MAF's transparent messages were disseminated through the media.

MAF did have a crisis management plan and was able to test it during Exercise Taurus, a simulation training exercise. Exercise Taurus proved useful for MAF and NZFSA even though the communication networks had not been tested. While the simulation training enabled the agencies to see strong and weak points in their responses, MAF and NZFSA would have learned more if these communication networks had been tested.

MAF officials admitted they do not continuously update and evaluate their crisis plans; however, after Exercise Taurus and *Operation Waiheke*, the agency reviewed its crisis management and response plan (Primary Production Committee, 2006). *Operation Waiheke* provides an opportunity and motivation for other organizations and government agencies to vicariously learn about and plan for the possibility of hoax events. It is important for organizations to continuously update their current plans and also to consider response planning to hoax events.

MAF and NZFSA found that establishing networks with stakeholders before the hoax was crucial during the response. Developing relationships with multiple stakeholders proved effective. The multiple networks provided additional resources to disseminate the information and reduce the uncertainty of the event. It cannot be stressed enough that organizations should reach out to their partners and other agencies before a crisis or hoax occurs.

MAF coordinated networks and formed partnerships before the hoax, but the agency was also able to create additional partnerships during the response. Despite MAF's crisis plan for the creation of a community liaison group, the agency knew it had to act quickly with the locals on the island to communicate the hoax response. Establishing the connection between the government and the local community chair provided useful ways for a two way flow of communication between the lead agency and the public. The public's role in the hoax response was two-fold. Initially, the public, particularly the farmers on Waiheke, were upset about the late notification of the situation. They did not think MAF's response was adequate. After MAF partnered with the community chair and held a public meeting, the farmers' view changed. The ability to listen to the public and engage them in a dialogue during the hoax response created a trusting environment between the government and the public.

The agencies (MAF, NZFSA and the NZ police) worked with the media to meet their press deadlines and scheduled regular media conferences. In turn, the media communicated the hoax message to the public. While it was helpful to work with the media's deadlines, MAF officials raised some concerns as their key decision makers were also their spokespeople.

MAF communicated with compassion and empathy–without over-reassuring those threatened by the risk of an outbreak. Due to the absence of foot and mouth symptoms and no loss of life, MAF's message focused on empathy and appreciation for people on Waiheke. This best practice stresses how a spokesperson should acknowledge the public's concerns and demonstrate human compassion, thereby being perceived by the public as a credible messenger.

Even though the hoax never developed into a crisis, MAF provided many messages of self-efficacy. Providing individuals with the tools to identify foot and mouth symptoms was one part of MAF's hoax response. Even though no cases of foot and mouth disease were identified, sharing the symptoms was an effective means of providing self-efficacy.

References

Action on second threat outside Waiheke could hurt trade. (2005, May 13). *Financial Times*. Retrieved December 6, 2006 from LexisNexis Academic database.

All signs point to letter hoax. (May 14, 2005). *The Dominion Post*. Retrieved December 6, 2006, from LexisNexis Academic database.

Anti-Hoax Terrorism Act of 2003: Hearing before the Subcommittee on Crime, Terrorism and Homeland Security of the Committee on the Judiciary House of Representatives, 108 Cong., 44. (2003). Retrieved July 2006 from http://commdocs.house.gov/committees/judiciary/hju88205.000/hju88205_0f.htm.

Biosecurity New Zealand. (n.d.) *Briefing for incoming ministers*. Retrieved September 30, 2005 from http://www.maf.govt.nz/mafnet/publications/2005-briefing-for-incoming-ministers/biosecurity-briefing.pdf.

Biosecurity threat investigation. (2005, May 10). Retrieved September 30, from www.maf.gov.nz/mafnet/press/100505fmd.htm.

Controlled area notice under the Biosecurity Act 1993. (2005, May 10). Retrieved September 30, from www.maf.gov.nz/mafnet/press/100505controlled-area-notice.htm.

Current Waiheke Island update. (2005, May 12). Retrieved September 30, from www.maf.gov.nz/mafnet/press/120505fmd2.htm.

D-day for foot and mouth threat. (2005). *The Dominion Post*. Retrieved December 6, 2006, from LexisNexis Academic database.

Disease alert may be capping stunt. (2005, May 11). *RNZ/One News*. Retrieved December 6, 2006, from LexisNexis Academic database.

Disease checks stay despite confidence. (2005, May 14). *The New Zealand Herald*. Retrieved December 6, 2006 from LexisNexis Academic database.

Farmers left in the dark. (2005, May 11). The *Dominion Post*. Retrieved December 6, 2006, from LexisNexis Academic database.

Ferries running normally to New Zealand "foot and mouth" island. (2005, May 10). *Radio New Zealand International*. Retrieved December 6, 2005 from Proquest Newspapers database.

Foot and mouth alert costing millions. (2005, May 12). *RNZ*. Retrieved from October 26, 2006 from http://tvnz.co/nz

Foot and mouth scare; Repercussions felt worldwide (2nd ed). (2005, May 12). *Dominion Post*, p. A5. Retrieved December 6, 2005, from Proquest Newspapers database.

Foot and mouth would have dire impact. (2005, May 11). *RNZ/TVNZ Interactive*. Retrieved October 26, 2006, from http://tvnz.co/nz

Hotere, A. (2005, May 15). Foot and mouth scare linked to test run. *The Sunday Star Times*. Retrieved December 6, 2006 from LexisNexis Academic database.

House of Representatives. (2001). *Anti-Hoax Terrorism Act of 2001* (Report 107–306). Washington, DC: US Government Printing Office.http://www.clerk.parliament.govt.nz/Publications/CommitteeReports

Kay, M. (2005a, May 11). Foot and mouth damage control. *The Press*, p. A1. Retrieved December 6, 2005, from Proquest Newspapers database.

Kay, M. (2005b, May 11). 'Hoax' a $10b threat to NZ. *Dominion Post*, p. A1. Retrieved December 6, 2005, from Proquest Newspapers database.

Kiwi shakes off disease hoax. (2005, May 11). *Bancorp Treasury*. Retrieved October 26, 2006, from http://tvnz.co/nz

Letter sparks foot and mouth scare. (2005, May 10). *One News/RNZ*. Retrieved October 26, 2006, from http://tvnz.co.nz.

Life on Waiheke like being in dream land. (2005, May 11). *Financial Times*. Retrieved December 6, 2006, from LexisNexis Academic database.

MAF braces for rash of foot and mouth scares in coming days. (2005, May 12). *Financial Times*. Retrieved December 6, 2006, from LexisNexis Academic database.

Media advisory: Media conference claimed foot and mouth threat. (2005a, May 11). Retrieved September 30, from www.maf.gov.nz/mafnet/press/110505fmd.htm

Media advisory: Latest media conference re: Waiheke Island situation. (2005b, May 13). Retrieved September 30, from www.maf.gov.nz/mafnet/press/130505fmd.htm

New Zealand Ministry of Agriculture and Forestry. (2006). *Exercise Taurus and Operation Waiheke: Lessons learned during a simulation exercise and a response suspected outbreak of foot-and-mouth disease*. (Reference number AX70-034-2). Wellington, NZ: Author.

NZ foot and mouth scare confirmed a hoax. (2005, May 25). *Xinhua News Agency*. Retrieved December 6, 2006, from LexisNexis Academic database.

NZ police pursue foot and mouth hoax clues. (2005, May 12). *Xinhua News Agency–CEIS Woodside*, p. 1. Retrieved December 6, 2005, from Proquest Newspapers database.

Office of MAF Director Post-clearance (Producer). (2005). *Collection of television news coverage of Operation Waiheke* [Video news clips]. (Available from MAF Post-clearance director, Wellington, NZ)

Operation Waiheke media advisory 2. (2005a, May 10). Retrieved September 30, from www.maf.gov.nz/mafnet/press/100505fmd3.htm

Operation Waiheke media advisory 3. (2005b, May 11). Retrieved September 30, from www.maf.gov.nz/mafnet/press/110505fmd2.htm

Operation Waiheke media advisory. (2005c, May 11). Retrieved September 30, from www.maf.gov.nz/mafnet/press/110505fmd3.htm

Primary Production Committee (March 8, 2006). Report of Exercise Taurus and *Operation Waiheke* Report of the Primary Production Committee, New Zealand House of Representatives. Retrieved March 12, 2006, from http://www.parliament.nz/en-NZ/SC/Reports.

Quick action on foot and mouth threat. (2005, May 11). *RNZ/One News*. Retrieved October 26, 2006, from http://tvnz.co/nz.

Roundup: Foot and mouth scare eases in New Zealand by Xia Wenhui. (2005, May 13). *Xinhua News Agency – CEIS Woodside*, p. 1. Retrieved December 6, 2005, from Proquest Newspapers database.

Scare set to hit meat trade. (2005, May 11). *The New Zealand Herald*. Retrieved December 6, 2006, from LexisNexis Academic database.

Second letters says foot and mouth threat a hoax. (2005, May 16). Retrieved September 30, from www.maf.gov.nz/mafnet/press/160505fmd.htm

Sell, B., and Berry, R. (2005, May 14). Day that put hearts in mouths. *The New Zealand Herald*. Retrieved October 17, 2006, from http://www.nzherald.co.nz

Stockfeed appeal for Waiheke farms. (2005, May 13). *The Evening Standard*. Retrieved December 6, 2006, from LexisNexis Academic database.

The temperature at ground zero. (2005, May 12). *The New Zealand Herald*. Retrieved December 6, 2006, from LexisNexis Academic database.

Ulmer, R. R., Sellnow, T. L., and Seeger, M.W. (2007). *Effective crisis communication: Moving from crisis to opportunity*. Thousand Oaks: Sage.

Waiheke farmers gear up for another round of stock testing. (2005, May 14). *The New Zealand Herald*. Retrieved October 17, 2006 from http://www.nzherald.co.nz

Chapter 8
Odwalla: The Long-Term Implications of Risk Communication

Written with Jennifer Reierson

On October 30, 1996, Odwalla, a company catering to health and nutrition conscious consumers, was alerted to a link between its apple juice and an *E. coli* outbreak. Contaminated apple juice has devastating potential given regular consumption by young children. Although the outbreak led to the death of an infant and illness of many other children, the majority of Odwalla's publics and stakeholders believed in its response and supported the company. The company, although struggling slightly after the crisis, underwent what many hailed as a successful renewal (Auerswald, 1999; Griffin, 2002; Odwalla, Inc., 1997). Odwalla retained 80% of its accounts (Odwalla, Inc.) and affected families and customers stood behind Odwalla as the company established itself as an authority on industry safety.

The Odwalla case reveals how one company can embrace a crisis and use it as an opportunity to become an industry leader, initiate industry wide change, and encourage organizational renewal. Odwalla's crisis happened at a vulnerable time for the organic food industry. As a result, the entire industry was transformed by stricter regulations and new and alternative production processes. This chapter reveals how Odwalla effectively used best practices in risk communication to learn and renew itself as well as the industry in which it operates. This case proceeds as follows: (1) an overview of the case, including a time line, is given; (2) the case is then analyzed using the best practices in risk communication; and finally (3) practical implications for effective risk communication identified from this case are provided.

Managing Risk Communication During an *E. Coli* Outbreak at Odwalla

On October 16, 1996, Children's Hospital in Denver admitted a 16-month old girl, Anna Gimmestad, with *E. coli* poisoning (Thomsen & Rawson, 1998). In late October 1996, the Food and Drug Administration (FDA) received information from three different state health departments that there was an *E. coli 0157:H7* outbreak (Fig. 8.1). Subsequent word from Washington State linked the outbreak to Odwalla

T. L. Sellnow et al. *Effective Risk Communication*

DOI: 10.1007/978-0-387-79727-4_8

apple juice (Henkel, 1999). On October 30, 1996, Odwalla was alerted by Washington's Department of Public Health of a link between their apple juice and a number of *E. coli* cases (Evan, 1999; Martinelli & Briggs, 1998). Odwalla immediately initiated a voluntary recall of all products containing apple juice. On October 31, 1996, Odwalla hired Edelman Public Relations' San Francisco office to take over the crisis response and communication (Evan). The same day Odwalla hired Edelman, the FDA launched a 14-month investigation (Henkel).

Odwalla expanded the voluntary recall on November 1 to include carrot and vegetable juice products, 13 juices in total (Evan, 1999; Martinell & Briggs, 1998). Delivery trucks were dispatched to retrieve the juice from retailers in seven states; internal task teams completed the recall in 48 hours ("Companies in crisis," 2005). Odwalla provided press release November 2 stating the "voluntary recall" was complete (Thomsen & Rawson, 1998).

Within 24 hours Odwalla had an explanatory website ("Companies in crisis," 2005). Odwalla also announced refunds to anyone who had purchased the juice and payment of medical costs for illness resulting from juice consumption (Martinelli & Briggs, 1998). On November 3, Odwalla's executives defended their sanitation process in Seattle newspapers and launched a second website to offer an alternative source for information about the recall–in addition to two 800-numbers previously created for customers and suppliers respectively (Martinelli & Briggs, 1998).

On November 4, 1996, the FDA confirmed that Odwalla's apple juice was responsible for the *E. coli* outbreak (Evan, 1999; Thomsen & Rawson, 1998). Odwalla explained that samples of apple juice from Odwalla's Tukwila, Washington, distribution center were tested and confirmed by the company to be infected with *E. coli* (Thomsen & Rawson, 1998). In yet another press conference held November 5, Steltenpohl suggested the company might look at heat pasteurization as a method for making apple juice in the future (Martinelli & Briggs, 1996; Thomsen & Rawson, 1998). He also indicated that he visited personally with the families of several of the affected children.

Tragically, the crisis took its first (and only) life on November 8, when Anna Gimmestad died from complications of *E. coli* poisoning contracted from Odwalla apple juice (Thomsen & Rawson, 1998). Following her death, Odwalla issued a press release offering condolences to her family. Another press release on November 8 indicated that no *E. coli* bacteria were found at Odwalla's Dinuba plant.

The focus shifted to suppliers and unsafe actions, as well as fresh-juice industry processes (Thomsen & Rawson, 1998). Safeway, a large grocery store chain, turned away un-pasteurized juice of Odwalla's competitors, adding to the tense situation between Odwalla and other organic juice companies. Stephen Williamson, Odwalla CEO, announced the company was forming a team of experts to define and generate solutions for the problem (Evan, 1999; Martinelli & Briggs, 1998). According to Martinelli and Briggs (1998), November 19 media reports said (1) Odwalla was calling on its competitors to stop selling un-pasteurized apple juice; (2) Odwalla was considering some type of heat treatment process for its apple juice; and (3) Odwalla announced that after having spent 11 days at its plant in Dinuba, California, the Food and Drug Administration reported it had found no evidence of *E. coli* bacteria.

The FDA did, however, point out concerns at the plant. Specifically, production process did not follow recommended federal guidelines for good manufacturing processes.

Large newspaper ads on December 5 announced Odwalla's use of a new process called *flash pasteurization* (Martinelli & Briggs, 1998). Ultimately, Odwalla determined flash pasteurization was the most effective guard against *E. coli*, while maintaining as much flavor and nutrients as possible. The juice went back on shelves and Odwalla looked to the future. Although sixty sales and distribution employees were laid off, the specific financial cost of the crisis to the company was not released (Martinelli & Briggs). The crisis affected not only Odwalla's image and reputation, but also the risk management procedures for the entire organic industry. In just one month, from October 30, 1996, to December 5, 1996, (see Figure 8.1) a company at the top of the organic juice industry saw the onset of an *E. coli* crisis as a result of its contaminated product, the infection of some sixty plus people (Griffin, 2002), the death of a child, the disruption of company vision and values, the plummeting of profits, and the overhaul of industry standards.

October 16, 1996	Anna admitted to Children's Hospital.
"Late October"	FDA receives word of *E. coli 0157:H7* outbreak.
October 30	Odwalla alerted to *E. coli* link and recalls apple juice.
October 31	Odwalla hires Edleman. FDA launches 14 month investigation.
November 1	Odwalla expands recall, announces refunds and payment of medical bills, launches website, sets up 1-800 numbers.
November 2	Voluntary recall complete.
November 3	Second website launched. Odwalla defends sanitation process in newspaper articles.
November 4	FDA confirms link to Odwalla. Tukwila, WA, apple juice samples Confirmed to contain *E. coli*.
November 5	Odwalla announces potential new heat pasteurization processes.
November 8	Anna Gimmestad dies. Odwalla announces no *E. coli* at Dinuba plant.
November 19	Media report: Odwalla calls on other companies to act. FDA found no *E. coli* at Odwalla's Dinuba plant. Odwalla considering heat treatment.
December 5	Odwalla announces use of flash pasteurization.

Fig. 8.1 Timeline

Applying the Best Practices in Risk Communication to the *E. coli* Outbreak at Odwalla

Lessons in risk communication reveal themselves throughout the stages of a crisis. Pre-crisis, crisis, and post-crisis actions and responses are interconnected and impacted by actions in previous stages. Consequently, each organizational action influences the potential for recovery and renewal. Although lines between these stages are not clearly delineated (Seeger et al., 2003) identifying stages can be helpful for analysis and planning.

Meet Risk Perception Needs by Remaining Open and Accessible to the Public

Odwalla maintained open communication with the media, held press conferences, updated a crisis specific website, instituted a hotline, and answered all calls from consumers and press. Odwalla's actions and response reinforced its commitment to the media and consumers as primary stakeholders. Actions of leaders also reinforced accessibility through visits with victims' families.

In 1996, the Internet was a new tool for risk communication. Odwalla embraced this tactic and developed a website designed to reach young media savvy customers of Odwalla during the crisis response (Berger, 2000). The website answered questions and provided links to key government agencies, such as the Centers for Disease Control and Prevention (CDC) and the Food and Drug Administration (FDA). The site was a timely way to reassure the public and to provide and manage essential information for various publics and the media. Creation of the website reinforced Odwalla's concern for its publics and stakeholders to have up-to-date, readily available, and detailed information that could ease concerns and provide direction for action.

Odwalla's Director of Marketing, Robin Joy, revealed that Odwalla did not have a risk or crisis plan at the time of the outbreak, but moved quickly, responded to all calls from consumers and press, developed an advisory board, and shared all information openly (Brewer, 1997). Joy described Odwalla's efforts to stay in touch with wary consumers through 800 numbers and a website that kept consumers and others up-to-date on safety procedures. Odwalla met the needs of the media and other stakeholders by remaining accessible, sharing information, and providing numerous options for information retrieval.

The novel idea of a website served another function. Odwalla's Chairman, Greg Steltenpohl, noted the website was an access point which created a sense of openness by the company (Rapaport, 1997). Steltenpohl reiterated the company's primary concern was its customers' safety and health (Halsey, 1996). In a press briefing on November 18, 1996, Odwalla's Steltenpohl said:

> Our core values are based around the idea of optimal nutrition . . . It may be possible to heat treat and still maintain the primary nutritional content of our apple juice. We're researching that now and when we have all the information we'll make a decision and share our finding with the industry and the public. ("FDA report," 1996, p. 1)

Odwalla appeared willing to be a leader, to share information with the industry and public, and to remain accessible throughout the outbreak and investigation. Moreover, this communication indicates Odwalla had nothing to hide and was continuing to support and extol its original company vision and values for optimal nutrition and health.

Present Risk Messages with Honesty

Customers were told to discard any Odwalla products containing apple juice (Halsey, 1996). Seattle's Department of Public Health Director, Dr. Alonzo Plough, warned the public not to drink Odwalla products containing apple juice and government agencies provided further information about protection from *E. coli* ("Bacterial ailment," 1996). In addition, Odwalla offered refunds to customers for any Odwalla apple juice products already purchased (Martinelli & Briggs, 1998). Refunds provided action orientation for consumers–return any Odwalla juice products.

Odwalla also acted to minimize the public's risk through corrective action. Upon word of a connection between the company's juice and an *E. coli 0157:H7* outbreak, Odwalla immediately issued a voluntary recall of all products containing apple juice ("Bacterial ailment," 1996; Burros, 1996; "Outbreak of," 1996; Richards, 1996a; U.S. Department, 1996). Odwalla also recalled carrot and vegetable juice as a precaution and offered the names of all thirteen juices being voluntarily recalled ("Bacterial ailment," 1996). The voluntary recall was instituted prior to confirmatory microbial tests (Burros, 1996), thus indicating a proactive response by Odwalla to mitigate risk to its customers. Also, by offering the names of all its juices, Odwalla gave the public necessary information to stay safe.

Messages of self-efficacy and corrective action were supplemented with messages that demonstrated Odwalla's dedication to reduce the risk of future contamination. Odwalla's CEO, Williamson, noted the need for safety to be the top priority (Groves, 1997). Williamson acknowledged the company's massive efforts and $1.5 million investment to upgrade safety and production measures. Odwalla hired a consultant who built a program to kill contaminants at every possible entry point, a systematic program called Hazards Analysis Critical Control Point plan (HACCP) (Drew & Belluck, 1998). Odwalla took proactive steps by investing significant amounts of money and time to determine infrastructure and process changes that would ensure the safety of its products.

Acknowledge Diverse Levels of Risk Tolerance

Odwalla's communication during the crisis reinforced the dissonance of the crisis with company vision. Odwalla executives apologized publicly for the outbreak and harm it caused (Moore, 1998). Company leaders acknowledged the tragedy of the situation and voiced the company's concern for victims and families (Gammon,

1996). Odwalla paid medical bills, visited patients, and compensated victims. Public relations specialists praised Odwalla's quick response, voluntary recall, cooperation with investigation, and payment of medical bills (Howe, 1996). Odwalla's Williamson acknowledged the crisis' impact on the company, "We have a scar and it's going to be here forever. It's important that we remember that, express our sorrow, and behave in a responsible way" ("New look," 1999, para. 12).

Company leaders reiterated concern for safety, but also acknowledged failure. Williamson said, "The bottom line is that you would have to say we failed" (Drew & Belluck, 1998, para. 23). Williamson said Odwalla now had "a more sensitive perspective on safety" and, "Having Anna die was unquestionably a horrifically awful, terrible thing to go through. You can't image worse" (Drew & Belluck, 1998, para. 8).

At a press conference on November 18, 1996, Odwalla indicated it had stopped production until the company could confirm if a heat treatment process would allow its products to maintain their nutritional value ("FDA report," 1996; Burros, 1996). Care for safety and quality was emphasized over company losses associated with a halt in production. Besides voluntary actions to mitigate harm, Odwalla and other officials communicated with the public about how they could reduce risk of potential harm.

Collaborate and Coordinate About Risk Communication with Credible Information Sources

One of the most frequent and intense themes in the risk communication was a focus on the organics industry, specifically, fresh and unpasteurized juice. This focus led to substantial effects on the industry. In a November 18, 1996, press conference, Steltenpohl said Odwalla vowed to lead the industry in solving the mystery behind the *E. coli* outbreak ("FDA report," 1996). During the same press conference, Odwalla encouraged other industry members to hold production of fresh apple juice products until results of heat treatment tests, conducted by Odwalla, were available (Burros, 1996; "FDA Report," 1996). Odwalla specifically called upon other industry members to take action.

Odwalla framed its *E. coli* outbreak as a "call to action" for the fresh apple juice industry as a whole ("FDA Report," 1996). Steltenpohl noted that adherence to industry standards were not enough to ensure safety of manufacturing practices and elimination of potential contamination by *E. coli*. Dr. Michael Doyle, a University of Georgia microbiologist, explained that the absence of dangerous pathogens cannot be assured by testing the final product ("FDA Report," 1996). These statements implicated an entire industry, created consumer fear, and disgruntled other industry members (Simmers, 1996).

Odwalla positioned itself as an industry leader and educator in safer juice production (Gallagher, 1997). To this point, Steltenpohl said, "By sharing the knowledge we've gathered with other juice makers, we seek to ensure a thriving and safe fresh juice industry" (Gallagher, 1997, para. 5). Additionally, Gallagher, Director of Communications for Odwalla wrote, "We have used our experience to help increase public knowledge and understanding of this crucial issue by cooperating with the government, industry and the news media" (Gallagher, 1998, para. 2). Further, Odwalla offered criticism of some 1,500 companies, mainly roadside stands and other fresh juice farmers, for not heeding warnings from Odwalla's crisis (Drew & Belluck, 1998).

Odwalla's case, as well as its risk communication, focused attention on new issues in the fresh juice industry. Christopher Gallagher, Director of Communications for Odwalla, was quoted, "It is widely known that the Odwalla case rewrote the rule books regarding food microbiology," particularly in regards to *E. coli's* ability to live in an acidic environment (Nax, 1998, para. 20). Odwalla moved food safety issues into the spotlight, a positive, according to many food safety experts (Rodriguez, 1998). The tragedy, Gallagher reinforced, was not just about Odwalla, but also about the industry and food safety (Gallagher, 1998). For Odwalla, focusing on the industry shifted attention away from the crisis and permitted Odwalla to emerge as a leader in safety and product quality.

Infuse Risk Communication into Policy Decisions

In order to create an improved production process, Odwalla created the Odwalla Nourishment and Food Safety Advisory Council ("FDA report," 1996). Soon after the crisis, Odwalla had one of the most thorough bacteria testing programs available, increased washing procedures, and had decided to pasteurize its juice (Drew & Belluck, 1998). The new pasteurization process was called flash-pasteurization, a process previously foreign to the fresh juice industry ("Juice maker," 1996). Odwalla broke new ground in the industry, took proactive steps toward improvement, and set new standards for others in the industry to follow.

Moreover, Odwalla had developed an HACCP program, the first ever of its kind in the fresh apple juice industry, under the guidance of Dr. Nickelson and Dr. Michael Doyle. This plan brought what was a regulatory standard in the poultry, seafood, and beef industries to the fresh juice industry ("FDA report," 1996). Odwalla's Vice President of Quality Assurance, Linda Frelka, said of the company's HACCP plan, that although the company had been called paranoid, Odwalla would take no risks, "We need to know that we're producing a safe product" (Postlewaite, 1999, para. 4). Odwalla, therefore, set another standard for the industry by presenting an option that could be called upon by the publics or government to be enacted by all industry members.

Odwalla also created a five-point plan for quality assurance. The plan included (1) validating HACCP, (2) juice testing for every batch of fresh fruit, (3) environment testing daily at microbiological site tests, (4) flash pasteurization for all juices, and (5) independent audits (Williamson, 2000). Odwalla changed processes and procedures to regain public support and avoid future threat of contamination.

Odwalla's industry influence caused tension with other producers and industry members and made it difficult to escape the effects of Odwalla's crisis. As a result of Odwalla's crisis, new regulation and industry standards of risk management were developed for the fresh juice industry. *E. coli* was not uncommon in the food industry, but the situation with Odwalla was unprecedented. Harmful bacteria in fresh fruits and vegetables was a new concept revealed by Odwalla's crisis (Drew & Belluck, 1998). Rapid expansion of the natural food industry was noted as a potential concern by the FDA that required more oversight following Odwalla's *E. coli* outbreak (Richards, 1996b).

Natural food trade association officials, representing nine different associations, met at FDA headquarters in Washington, DC, within hours of word of the Odwalla *E. coli* outbreak (Richards, 1996b). Industry safety processes and procedures were called into question. Post-crisis stories emphasized pasteurization, a process most fresh juice producers did not use, was the best way to kill *E. coli* ("Questions of," 1996). Consequently, health officials revisited the idea of requiring pasteurization for widely distributed juices ("Questions of," 1996). Chemical washing for produce also came under debate by U.S. government officials. Again, officials discussed new requirements for the entire industry. Such changes, if implemented, would alter the entire purpose and belief system behind the fresh juice industry.

Following Odwalla's crisis, food scientists and health officials also debated how to prevent such crises. Some officials proposed stricter inspection requirements while others advised using the same process used for milk, requiring pasteurization for any product transported across state lines ("Question of," 1996). The concept of required pasteurization raised questions concerning appropriate levels of regulation. No previous evidence of widespread contamination from fruit juices existed, thus Federal regulators had not addressed the issue of pasteurization for fresh juices. Odwalla's crisis in effect became a public health and policy issue. Odwalla's scenario opened the door to an entirely new type of crisis for the fresh juice industry and a new point of potential regulation for Federal agencies.

By January 1997, the FDA was encouraging greater vigilance from regulators, and new rules regarding pasteurization were being drafted, but not without opposition from producers standing by the better taste of unpasteurized products (Drew & Belluck, 1998). Federal regulators drafted rules for the juice industry requiring an HACCP plan. The FDA also proposed that by year 2000 all fresh juices were required to either be pasteurized or undergo alternative processes to remove harmful bacteria (Burros, 1998). In addition, the FDA proposed a warning label for unpasteurized juices that read, "may contain harmful bacteria which can cause serious illness in children, the elderly, and persons with weakened immune systems" ("Warning label," 1998, para. 4). Odwalla's 1996 *E. coli* outbreak was cited as the reason for the new label and also for the requirement of new steps in the juice treatment process.

Some apple growers, for instance those in El Dorado County, made changes to reassure customers after the Odwalla *E. coli* outbreak (Lifsher, 1997). The group of 40 growers, called Apple Hill growers, developed a quality assurance plan because, they complained, pasteurization would be too expensive and ruin the taste of their juices. The HACCP plan provided stringent guidelines and required submission to government inspections in order to obtain the right to display the plan's logo.

Odwalla also set a new standard for criminal charges in the food industry. Federal officials brought criminal charges against Odwalla in an attempt to expand penalties that would compel food supply companies to create increasingly rigorous standards for safety (Belluck, 1998). Criminal charges were not common in food-safety cases and few criminal laws existed that addressed food-safety.

Odwalla's *E. coli* crisis created an atmosphere ripe for fulfilling lessons of process approach and policy development. Odwalla embraced the process approach by reevaluating, changing, and improving standards, practices, and policies. Odwalla also realized opportunities for leadership that helped the company move through and out of the crisis. However, this crisis had a profound effect not only on Odwalla but also the fresh juice industry and the organic food industry. Food safety issues were scrutinized, and new regulations, laws, and legal precedents were developed. Growers and producers at all levels felt the effects of Odwalla's crisis. Consequently, Odwalla's crisis revealed lessons in industry implications and tensions.

Implications for Effective Risk Communication

The Importance of Consistent Ethical Behaviors Pre-Crisis

Within the crisis and post-crisis stages, Odwalla's pre-crisis actions were questioned. A lapse of consumer-oriented behavior prior to the crisis became problematic as accusations of use of rotten fruit, a stressed production system, disputes about safety testing, and rushed production surfaced (Drew & Belluck, 1998). As a result, Odwalla was forced to address these accusations, thereby restricting its ability to focus exclusively on moving forward. Unsafe practices prior to a crisis can restrict and slow post-crisis progress and renewal. Conversely, enacting and underscoring safe and ethical practices prior to a crisis can facilitate positive and forward-looking post-crisis communication.

The Importance of Quick Action in Risk Communication

Odwalla realized the need for quick action and communication following word of a potential link between *E. coli* and its apple juice. Odwalla set a strong example of proactive and voluntary actions to pull products, stop production, and set up information sources such as hotlines and websites. These actions took place before confirmatory tests were completed. In this way, Odwalla reinforced its credibility and legitimacy as a company concerned with health and safety. Further, Odwalla's quick actions may have mitigated or prevented the risk of potential harm to other customers. Company's faced with food safety concerns should act quickly, even before confirmatory tests, in order to maintain credibility, limit harm, and portray an image of care and concern. Maintaining legitimacy and credibility in the long run may offset initial costs associated with halted production or product recall.

The Importance of Concern and Empathy

Odwalla also used the media in an effective manner to show concern and empathy, provide the public with crucial information, and raise awareness about efforts to prevent future contamination and correct problems with the current process. Being open and honest and easily accessible to the media afforded Odwalla respect that allowed the company to expound messages to the public. Companies in crisis should follow Odwalla's lead and provide the media and public with information about what is being done to correct the problem, reduce the risk to publics, and prevent future crises.

Moreover, Odwalla reinforced the company's vision through expressed remorse and heartfelt concern for families affected by the *E. coli* poisoning. Company leaders became real, caring individuals who grieved along with the families. Through consistent responses, based on ethics, vision, and strong values, an organization can respond to complexities of crisis in a way that establishes and reinforces dedication to stakeholders and subsequently, renewal. Companies facing food safety crisis should not forget the human element and should respond to and recognize victims. Further, a company in crisis should revisit its vision and values and ensure that its response is consistent with these.

The Importance of Ethical Communication and Resources

During the response, Odwalla's primary concern became the safety of its products, customers, and production process. Worries about lost profit and stocks were absent as Odwalla spent money to do what they thought was responsible, ethical, and in accordance with the company vision. For example, Odwalla spent money on victim's hospital bills, communicating messages effectively, consulting industry and safety experts, and changing production processes and standards. Other companies should do what is financially possible to behave in similar ways. However, from this we learn that in order for a company to renew and adhere to some of the best practices, significant resources may be needed.

Odwalla was fortunate enough to be able to define the company's path during post-crisis as leading the industry to new, safer processes and procedures. Odwalla had resources available for this cause such as a $27 million insurance policy to help cover legal fees (Castaneda, 1997; Veverka, 1996). Further, Odwalla had reserves of $10 million in cash (Veverka, 1996). As a company that brought in $59.2 million in fiscal 1996 (Veverka, 1996), Odwalla was positioned financially to withstand a costly crisis. Resources were needed to deal with litigation, settlements, loss of sales, public relations responses, and corrective actions. However, smaller, less profitable companies may not have been able to define a similar path. Therefore, without significant monetary resources, an organization will likely find renewal and consistent adherence to best practices difficult to achieve.

The Importance of Organizational Learning and Industry Consideration

Finally, Odwalla's crisis response shifted focus to and mobilized an entire industry. One organization's crisis can damage the reputation of an entire industry and can also spur changes in risk management for an entire industry. An organization in crisis has the potential to speak for an entire industry and therefore should behave and respond in ways that acknowledge and support other industry members. Odwalla's goal to lead the industry in new safety standards and procedures stimulated comprehensive industry changes. Survival of other industry members could depend on whether or not a company in crisis considers widespread ramifications for the industry and acts in ethical and responsible ways. An organization in crisis should include other industry members in decisions about revised standards and procedures.

Odwalla's post-crisis communication focused on the natural juice industry and the ineffectiveness of common safety standards and techniques for killing bacteria and other pathogens. According to Odwalla, its crisis underscored the need for updated industry procedures and practices. Implied, then, is that companies not heeding the warning were acting irresponsibly. As a result of Odwalla's crisis, regulatory changes forced industry members to conform and implement costly modifications. Those that did not immediately change were eventually forced to pasteurize or include warning labels on their products. A lesson learned, then, is that organizations in times of shifting risk management standards should endeavor to nurture an environment of collaboration with and responsibly represent other industry members.

Damage and harm can be widespread and quick in any crisis. Therefore, quick reaction and response are paramount. Moreover, adherence to lessons learned from previous crises and recognition of opportunities to respond in ethical and positive ways are important steps in risk management and facilitation of renewal. Analysis of Odwalla's case provided the opportunity for valuable lessons to emerge. The goal of this analysis and similar analyses is that organizations, leaders, regulatory agencies, and industry personnel learn vicariously from the experience of others.

References

Auerswald, B. A. (1999, June/July). Restocking the shelves: Recovering from a recall. *Food Quality Journal*. Retrieved July 11, 2006, from http://www.psandman.com/articles/restock.htm.

Bacterial ailment traced to fruit juice brand (1996, November 1). *The New York Times*, p. A28. Retrieved September 5, 2006, from ProQuest Newspapers, ABI/Inform Global databases: http://proquest.umi.com/pqdweb?RQT=302&cfc=1.

Belluck, P. (1998, July 24). Juice-poisoning case brings guilty plea and a huge fine. *The New York Times*, p. A12. Retrieved September 5, 2006, from ProQuest Newspapers, ABI/Inform Global databases: http://proquest.umi.com/pqdweb?RQT=302&cfc=1.

Berger, W. (2000, October 25). Spinner's web weapons: T-chips and dark sites. *The New York Times*, p. H32. Retrieved September 5, 2006, from ProQuest Newspapers, ABI/Inform Global databases: http://proquest.umi.com/pqdweb?RQT=302&cfc=1.

Brewer, G. (1997). When a crisis squeezes sales. *Sales and Marketing Management, 149*(6), 95. Retrieved September 5, 2006, from ProQuest Newspapers, ABI/Inform Global databases: http://proquest.umi.com/pqdweb?RQT=302&cfc=1.

Burros, M. (1996, November 20). Opting for an early warning when *E. coli* is suspected. *The Wall Street Journal*, p. C3+. Retrieved September 5, 2006, from ProQuest Newspapers, ABI/Inform Global databases: http://proquest.umi.com/pqdweb?RQT=302&cfc=1.

Burros, M. (1998, July 1). Is it pasteurized? Juice labels will tell. *The New York Times*, p. F5. Retrieved September 5, 2006, from ProQuest Newspapers, ABI/Inform Global databases: http://proquest.umi.com/pqdweb?RQT=302&cfc=1.

Castaneda, L. (1997, April 9). Odwalla struggling back from recall. *San Francisco Chronicle*, p. D2. Retrieved September 5, 2006, from ProQuest Newspapers, ABI/Inform Global databases: http://proquest.umi.com/pqdweb?RQT=302&cfc=1.

Companies in crisis—what to do when it all goes wrong: Odwalla and the E-coli outbreak. (2005). *Mallenbaker.net*. Retrieved January 17, 2006, from http://www.mallenbaker.net/csr/CSRfiles/crisis05.html.

Drew, C., & Belluck, P. (1998, January 4). Deadly bacteria a new threat to fruit and produce in U.S.: Fresh hazards [first of three articles]. *The New York Times*, p. 1.1. Retrieved September 5, 2006, from ProQuest Newspapers, ABI/Inform Global databases: http://proquest.umi.com/pqdweb?RQT=302&cfc=1.

Evan, T. (1999). Odwalla. *Public Relations Quarterly, 44*, 15–17.

FDA report indicates no *E. coli 0157:H7* found at Dinuba plant. (1996, November 23). *KidSource Online*. Retrieved January 17, 2006, from http://kidsource.com/kidsource/content2/ecoli/odwalla.11.23.html.

Gallagher, C. (1997, October 13). Odwalla's 100% pure pressed carrot juice returns. *PR Newswire*, p. 1. Retrieved September 5, 2006, from ProQuest Newspapers, ABI/Inform Global databases: http://proquest.umi.com/pqdweb?RQT=302& cfc=1.

Gallagher, C. C. (1998, January 9). Company took steps [Letter to the editor]. *The New York Times*, p. A18. Retrieved September 5, 2006, from ProQuest Newspapers, ABI/Inform Global databases: http://proquest.umi.com/pqdweb?RQT=302&cfc=1.

Gammon, R. (1996, October 31). Odwalla recalls apple drinks. *Santa Cruz Sentinel*, p. A1. Retrieved September 5, 2006, from ProQuest Newspapers, ABI/Inform Global databases: http://proquest.umi.com/pqdweb?RQT=302&cfc=1.

Griffin, G. (2002, August 4). ConAgra lies low after recall: Experts pan lack of public reassurance. *Denver Post*. Retrieved July 11, 2006, from http://www.petersandman.com/articles/conagra.htm.

Groves, M. (1997, October 30). News and insight on business in the golden state; A maturing experience; Last year's *E. coli* outbreak forced juice maker to learn a lot in a hurry. *Los Angeles Times*, p. 2. Retrieved September 5, 2006, from ProQuest Newspapers, ABI/Inform Global databases: http://proquest.umi.com/pqdweb?RQT=302&cfc=1.

Halsey, E. (1996, November 1). *E. coli* poisoning leads to Odwalla juice recall. *CNN Interactive*. Retrieved January 17, 2006, from http://www.cnn.com/HEALTH/9611/01/e.coli.poisoning/.

Henkel, J. (1999). Juice maker fined record amount for *E. coli*-tainted product. *FDA Consumer, 33*, 34.

Howe, K. (1996, November 2). Odwalla gets high marks for concern. *San Francisco Chronicle*, p. D1. Retrieved September 5, 2006, from ProQuest Newspapers, ABI/Inform Global databases: http://proquest.umi.com/pqdweb?RQT=302&cfc=1.

Juice maker to pasteurize. (1996, December 6). *The New York Times*, p. A20. Retrieved September 5, 2006, from ProQuest Newspapers, ABI/Inform Global databases: http://proquest.umi.com/pqdweb?RQT=302&cfc=1.

Lifsher, M. (1997, September 17). Apple growers revamp to reassure wary public. *The Wall Street Journal*, p. CA.1. Retrieved September 5, 2006, from ProQuestNewspapers, ABI/Inform Global databases: http://proquest.umi.com/pqdweb?RQT=302&cfc=1.

Martinelli, K., & Briggs, W. (1998). Integrating public relations and legal responses during a crisis: The case of Odwalla, Inc. *Public Relations Review, 24*, 443–460.

Moore, B. L. (1998, August 19). Time may be right to take bite of Odwalla. *The Wall Street Journal*, p. CA 1. Retrieved October 26, 2006, from ProQuest Newspapers, ABI/Inform Global databases: http://proquest.umi.com/pqdweb?RQT=302&cfc=1.

Nax, S. (1998, May 27). Odwalla agrees to settle 5 lawsuits: The agreement resolves 17 of 20 lawsuits filed against the juice maker over an outbreak of E. coli bacteria in 1996. *The Fresno Bee*, p. C1. Retrieved September 5, 2006, from ProQuest Newspapers, ABI/Inform Global databases: http://proquest.umi.com/pqdweb?RQT=302&cfc=1.

New look for Odwalla juices helps to mend image. (1999, September 20). *The Wall Street Journal*, p. B.11.G. Retrieved September 5, 2006, from ProQuest Newspapers, ABI/Inform Global databases: http://proquest.umi.com/pqdweb?RQT=302&cfc=1.

Odwalla, Inc. (1997, August 22). *The Wall Street Journal*, p. A6. Retrieved September 5, 2006, from ProQuest Newspapers, ABI/Inform Global databases: http://proquest.umi.com/pqdweb?RQT=302&cfc=1.

Outbreak of *Escherichia coli 0157:H7* infections associated with drinking unpasteurized apple juice. (1996). *Journal of the American Medical Association, 276*, 1865.

Postlewaite, K. (1999). Critical conditions. *Beverage Industry, 90(3)*, 26–27. Retrieved September 5, 2006, from ProQuest Newspapers, ABI/Inform Global databases: http://proquest.umi.com/pqdweb?RQT=302&cfc=1.

Questions of pasteurization raised after *E. coli* is traced to juice. (1996, November 4). *The New York Times*, p. A17. Retrieved September 5, 2006, from ProQuest Newspapers, ABI/Inform Global databases: http://proquest.umi.com/pqdweb?RQT=302&cfc=1.

Rapaport, R. (1997, October 6). PR finds a cool new tool. *Forbes*, p. 100. Retrieved September 5, 2006, from ProQuest Newspapers, ABI/Inform Global databases: http://proquest.umi.com/pqdweb?RQT=302&cfc=1.

Richards, B. (1996a, November 1). Odwalla's contaminated apple juice blamed for *E. coli* outbreak in Seattle. *The Wall Street Journal*, p. B3+. Retrieved September 5, 2006, from ProQuest Newspapers, ABI/Inform Global databases: http://proquest.umi.com/pqdweb?RQT=302&cfc=1.

Richards, B. (1996b, November 4). Odwalla's woes are a lesson for natural-food industry: FDA seeks tighter quality controls as *E. coli* raises concerns. *The Wall Street Journal*, p. B4. Retrieved September 5, 2006, from ProQuest Newspapers, ABI/Inform Global databases: http://proquest.umi.com/pqdweb?RQT=302&cfc=1.

Rodriguez, R. (1998, July 24). Odwalla plea puts spotlight on safety. *The Fresno Bee*, p. C1. Retrieved September 5, 2006, from ProQuest Newspapers, ABI/Inform Global databases: http://proquest.umi.com/pqdweb?RQT=302&cfc=1.

Seeger, M. W., Sellnow, T. L., & Ulmer, R. R. (2003). *Communication and organizational crisis*. Westport, CT: Praeger.

Simmers, T. (1996, November 23). Safeway bars Odwalla rivals from its shelves. *San Mateo Times*, p. C1. Retrieved September 5, 2006, from ProQuest Newspapers, ABI/Inform Global databases: http://proquest.umi.com/pqdweb?RQT=302&cfc=1.

Thomsen, S., & Rawson, B. (1998). Purifying a tainted corporate image: Odwalla's response to an *E. coli* poisoning. *Public Relations Quarterly, 43*, 35–46.

U.S. Department of Health and Human Services. (1996, October 31). *E. coli 0157: H7 outbreak associated with Odwalla brand apple juice products*. Retrieved January 1, 2006, from http://vm.cfsan.fda.gov/~lrd/apple.html.

Veverka, M. (1996, December 4). Odwalla still faces some problems that could make it a risky bet. *The Wall Street Journal*, p. CA.2. Retrieved September 5, 2006, from ProQuest Newspapers, ABI/Inform Global databases: http://proquest.umi.com/pqdweb?RQT=302&cfc=1.

Warning label proposed for unpasteurized juice. (1998, April 22). *The New York Times*, p. A16. Retrieved September 5, 2006, from ProQuest Newspapers, ABI/Inform Global databases: http://proquest.umi.com/pqdweb?RQT=302&cfc=1.

Williamson, S. (2000). Profiles in leadership: The big squeeze. *Risk Management, 47(9)*, 14–16. Retrieved September 5, 2006, from ProQuest Newspapers, ABI/Inform Global databases: http://proquest.umi.com/pqdweb?RQT=302&cfc=1.

Chapter 9
ConAgra: Audience Complexity in Risk Communication

Written with Elizabeth Petrun

In October 2007, the Centers for Disease Control and Prevention (CDC) announced than ConAgra pot pies, including Banquet and private label brands produced at a single ConAgra plant in Marshall, Missouri, were linked to nearly 300 cases of unique types of *Salmonella* (ConAgra, 2007). Within a short period of time, ConAgra Foods issued a recall of all brands associated with their pot pie product. Focusing on ConAgra's crisis, this case study provides a vivid demonstration of the complexity of risk communication during a recall event. It underscores the importance of addressing the multiple audiences associated with a product. This case study proceeds with an overview and timeline of the case, an examination of the messages issued by ConAgra using the conceptual framework of the best practices of risk communication, and practical implications for effective risk communication. This case proceeds as follows: (1) an overview of the case, including a time line, is given; (2) the case is then analyzed using the best practices in risk communication; and finally (3) practical implications for effective risk communication identified from this case are provided.

Managing Risk Communication During a *Salmonella* Outbreak

Reports to State Public Health Agencies first suggested an outbreak of *Salmonella* linked to the Banquet pot pies on October 4, 2007. A documented conference call with the CDC and the United States Department of Agriculture (USDA) followed on October 5 (National Center for Food Protection and Defense, 2007). On October 9, 2007, both the CDC and the Food Safety and Inspection Service (FSIS) issued public health alerts to warn consumers not to eat affected pot pie brands (National Center for Food Protection and Defense, 2007). Initially, ConAgra Foods sent an advisory specifically pertaining to "Banquet brand frozen chicken or turkey pot pie products or generic store brand not-ready-to-eat pot pie products bearing the number 'P-9' printed on the side of the package" (ConAgra Foods, n.d.). The company

T. L. Sellnow et al. *Effective Risk Communication*

DOI: 10.1007/978-0-387-79727-4_9

"directed retailers to remove the poultry pot pies from shelves, [and] suspended pot pie production in its Marshall, Missouri, plant" (ConAgra Foods, n.d.).

In a press release, ConAgra Foods informed their customers that, if they wanted to, they could return the product to ConAgra Foods "for a refund by sending the side panel of the package that contains the code 'P-9' to ConAgra Foods, Dept. BQPP, P.O. Box 3768, Omaha, NE 68103-0768" (ConAgra Foods, n.d.). The company also informed consumers that they could "return the product to the store from which it was purchased for a refund" (ConAgra Foods, n.d.). In their first press release, ConAgra Foods stated, "The Company believes the issue is likely related to consumer undercooking of the product" (ConAgra Foods, n.d.). ConAgra immediately began working with the USDA to define possible changes to the cooking instructions that needed to be made. The Marshall plant was closed on October 9, and ConAgra issued a market withdrawal. On October 11, 2007, a voluntary recall was issued.

In addition to the recall, ConAgra reminded consumers that the products were not cooked in advance and that it was working on redeveloping cooking instructions (ConAgra Foods, n.d.). The investigation into the source of the outbreak is still on-going. Currently, there is no new publicized information. While no deaths were reported due to the *Salmonella* outbreak, a minimum of 272 isolates were collected from ill individuals in 35 states (National Center for Food Protection and Defense, 2007). Ages of the infected ranged from less than 1 year to 89 years, and the median age was 18 (National Center for Food Protection and Defense, 2007). At least 65 people were hospitalized due to illness (National Center for Food Protection and Defense, 2007) (Fig. 9.1).

October 4, 2007	Suggested outbreak in Banquet pot pies
October 5, 2007	Conference call to the CDC and USDA
October 9, 2007	Marshall plant closed
October 9, 2007	ConAgra issued a withdrawal from the market
October 11, 2007	ConAgra issues a voluntary recall

Fig. 9.1 Timeline

When the USDA issued a health alert "to warn consumers of the link between ConAgra Foods' product and the *salmonella* cases" (Associated Press, 2007a, para. 2), all varieties of frozen pot pies linked to the *salmonellosis* outbreak were pulled from the market. This alert was issued to the entire population; technically, then, the general public was informed. However, information about the recall suggested the presence of multiple elements within this general audience, including the following:

- those who knew or did not know the meaning of the words "pot pies" or *salmonellosis*;
- those who had or had not purchased pot pies having one of the identified brands (Banquet, Albertson's, Food Lion, Great Value, Hill Country Fare, Kirkwood, Kroger, Meijer, Western Family);

- those who could or could not afford to discard any food items, even if contaminated;
- those who had or did not have access to media to get the establishment number "P-9" or identification information "Est. 1059" printed on the sides of recalled pot pies packages;
- those who were or were not in the states with illness outbreaks;
- those who were or were not in the especially threatened groups (infants, the elderly, persons with HIV infection or undergoing chemotherapy);
- those who did or did not experience the most common manifestations of *salmonellosis* (diarrhea, abdominal cramps, and fever within 8–72 hours; chills, headache, nausea and vomiting lasting up to 7 days);
- and those whose languages were or were not included on the toll-free USDA Meat and Poultry Hotline available in English and Spanish (e.g., Associated Press, 2007a; CIDRAP News, 2007; Eamich, 2007).

For industry and governmental spokespersons issuing warnings to the general public about a link between undercooked pot pies and *Salmonella,* the dilemma was that diverse elements within the general population may receive and process the information differently. The literacy level, economic status, access to media, proximity to the outbreak, membership in at-risk groups, health condition, and language proficiency of the listeners could affect the level of attention and compliance given to the message. This raises the question: What impact do cultural variables have on the sending and receiving of risk and crisis messages?

Applying the Best Practices of Risk Communication to the *Salmonella* Outbreak at ConAgra

The ConAgra case provides insight into a number of the best practices for risk communication. In this section, we emphasize the relationship of the message's complexity to the culture-centered needs of the audience. We also discuss best practices related to uncertainty, risk tolerance, maintaining honesty, openness, and collaboration.

Design Risk Messages to be Culture-Centered

A review of the risk and recall messages released by ConAgra and issued through the media suggests that little effort was made to recognize diverse perspectives or cultural groups. Initially, as reflected in sources drawn from October 10 through December 21, 2007, only general terms were used to describe the consumer. For example, in every message, general terms such as "customer," "people," "consumer," "patient," and "cases" were used when describing those who had been affected, hospitalized, or may have purchased the product. Specific terms were used only in reference to families suing because they believed *Salmonella* in ConAgra's pot pies made their children sick (Business Briefs, 2007, p. 13; Ruff, 2007b). The implication is that the communicators considered the general public, rather than particular publics, when constructing and issuing messages. This contributed to the perception that there was considerable distance between the company and the consumer in terms of empathy, compassion, or concern. In each of the following areas–literacy, economic status, access to the media, proximity to the outbreak, and at-risk groups– observations are offered reflecting the attention of the risk communicators to the general, rather than the particular audiences.

Literacy

There was a high expectation that consumers were able to read messages, identify codes, follow directions, and respond accordingly. As the situation unfolded, company spokespeople assumed that the consumers would understand the difference between risk alert and recall messages. The company initially suggested that "if their pies were…cooked properly" [Associated Press, 2007b, p. A.21] there should be no problem. The continued presence of the product within distribution centers may have contributed to this confusion. Michael Doyle, head of the University of Georgia's Center for Food Safety said, "Consumers might not be sufficiently aware that frozen meals need thorough cooking" (Ruff, 2007a). However, once the recall was issued, directions were quite explicit about what consumers should look for: "the number 'P-9' printed on the side of the package" (Schneider, 2007, p. B.5), ten-digit UPC Codes printed on the sides of packages ("Giant Food alerts," 2007), and address on the side panel of the package where consumers should send the code (Bratton, 2007; Weise & Schmit, 2007). The suggestion that consumers wanting a refund had recourse by sending proof of purchase to ConAgra Foods was often buried at the end of a news article or posted on the website. Attempting to read and thoroughly process this information tested the perseverance and ability of consumers with limited literacy skills. In addition, no language other than English was used to communicate about the recall or the contamination. No Hispanic or multilingual information sources were offered in any of the materials reviewed in this analysis. ConAgra's perception that only English-speaking people would have questions or concerns about the affected products was shortsighted.

Economic Status

Most of the media attention focused on the ConAgra's and its affiliates' perspective of how much the recall would cost them in terms of profit (Ruff, 2007c; Weise & Schmit, 2007). However, one specific reference to the value of the recalled products was made when an attorney was quoted as saying, "As of last night these products were still on store shelves and in fact were on sale—2 for $1.00. ConAgra, the USDA, and all health authorities should put people's safety above sales"

("William Marler," 2007, para. 1). The low cost of a pot pie provides a general indication of the lower economic status of most of those inclined to purchase this product.

The warning to customers that they should "not eat the pot pies" and "throw them away" may not have resonated with people living on fixed incomes or those with limited resources. The elderly, after living through the Great Depression, may have been prone to disregard the message. For families with hungry children, the prospect of having nothing to eat may have outweighed the risk of potential *Salmonella* poisoning.

Access to Media

Company spokespeople directed consumers to go to websites for additional information. For example, consumers were directed to The Food Safety and Inspection Service web site (Rosetta, 2007) for more information. However, the reality is that nearly one quarter of all Americans have no home computer access and the majority of these are situated in the lower socioeconomic strata. This suggests that the means of self-efficacy was directed toward those with media access, rather than those who may have been most affected by the *Salmonella* outbreak and product recall.

Proximity to Outbreak

Frequent mention was made to the locations and number of reported cases of *Salmonella* attributed to the consumption of ConAgra's Banquet pot pies (Rosetta, 2007; Ruff, 2007a, 2007c; "William Marler," 2007). The recall did not result in the immediate removal of the product from distribution centers. While the media provided information about where outbreaks of *Salmonella* had been attributed to the consumption of pot pies, the product remained on the shelves in various locations. The identification of the versions of the Banquet pot pies sold under the names of Albertson's, Hill Country Fare, Food Lion, Great Value, Kirkwood, Kroger, Meijer and Western Family (Bratton, 2007) provided some brand clarification.

Membership in at-Risk Groups

At-risk groups were identified as "people with weaker immune systems such as the elderly or very young" (Funk, 2007, p. A.1). These are the groups most likely to be eating pot pies, due to the economic reasons previously mentioned. Other potential consumers were not identified as being at risk. The symptoms of *Salmonella* were identified as diarrhea, fever, dehydration, abdominal pain and vomiting (Funk, 2007). These conditions are not unique to *Salmonella* poisoning and may have been perceived as worth the risk if weighed against having nothing to eat or wasting money.

Acknowledge Diverse Levels of Risk Tolerance

Risk and crisis messages are not always interpreted as intended. A variety of factors contribute to the misunderstanding of risk messages. When faced with developing messages of self-efficacy, sensitivity to stakeholder perceptions is essential. ConAgra Foods' initial response to the *Salmonella* outbreak did not include information of its interest for the public's concern pertaining to the contaminated product. Neither of the two initial news releases on ConAgra Foods' website discussed what *Salmonella* is and/or what the symptoms are. In addition, ConAgra did not provide any information to potential high-risk stakeholders, such as the elderly, the very young, and those with compromised immune systems. In both press releases, at the bottom of the page, there was a reference to go to International Food Information Council (IFIC) web site for information on food safety (ConAgra Foods, n.d.). Any time there is the potential for serious harm to stakeholder's health, stakeholders have the right to be overly informed about the threat faced. Organizations that realize a *Salmonella* outbreak has been connected to their products should provide clear messages of self efficacy and instruct the public about the levels of risk it may experience. They should not assume that everyone knows what *Salmonella* is and what possible effects it can have. Moreover, organizations should make ample attempts to inform the public and continue to be active in assessing the retention of their intended messages. In some emergency crisis situations, it is best to give stakeholders "simple tasks . . . [to] give people back a sense of control" (Reynolds, 2002, p. 24). Giving stakeholders something to do has other benefits, according to CDC risk expert Barbara Reynolds:

> [Providing tasks] will help to keep them motivated to stay tuned to what is happening (versus denial, where they refuse to acknowledge the possible danger to themselves and others) and prepare them to take action when directed to do so. When giving people something to do, give them a choice of actions matched to their level of concern. Offer a range of responses—a minimum response, a maximum response, and a recommended middle response. [For example,] To make drinking water safe: (1) Use chlorine drops if safety is uncertain, (2) boil water for 2 minutes or, (3) buy bottled water. (2002, p. 24)

Account for Uncertainty Inherent in Risk

Making statements that attempt to dispel uncertainty during a crisis situation is not the best approach. Research suggests that effective risk communication involves some level of "ambiguity or uncertainty that will enable the organization both to communicate with their public and to emphasize the level of uncertainty they are experiencing at the time" (Ulmer et al., 2007, p. 43). In short, an "organization must be able to communicate what it knows at the time" of a crisis (Ulmer et al., 2007, p. 43). This type of response will give the public a better sense of security.

On October 9, 2007, the day after it was informed by health officials that a number of consumers had been diagnosed with *Salmonella* poisoning and that the infection was believed to be linked to its frozen pot pies, ConAgra Foods decided to make an advisory to the public, specifically pertaining to its Banquet turkey and chicken pot pies (ConAgra Foods, n.d.). In its first press release, ConAgra stated that it believed the issue to be related to improper cooking by *consumers* (ConAgra Foods, n.d.). Working with the USDA, ConAgra Foods decided to revise the cooking instructions located on the food packages. ConAgra Foods' immediate response was to warn people about the pot pies containing poultry, because *Salmonella* is typically related to poultry. Instead of saying, "There have been a large number of people recently diagnosed with *Salmonella* poising and it is related to our pot pies," ConAgra told people to beware, *specifically*, of the Banquet turkey and chicken pot pies. While research has shown that organizations should "avoid certain or absolute answers to the public and media until sufficient information is available" (Ulmer et al., 2007, p. 43), ConAgra immediately shifted the blame to the customer, asserting that customers were not cooking the pot pies properly.

Risk and crisis communicators also should foster some level of uncertainty or ambiguity, because it "will enable them both to communicate with their public and to emphasize the level of uncertainty they are experiencing at the time–in effect, a more accurate reflection of the situation" (Ulmer et al., 2007, p. 43). ConAgra Foods' immediate response was stated as a fact. The first news release even assured consumers that it was working with the "USDA to identify any additional steps that may be appropriate, including potential changes that may further clarify cooking instructions for consumers" (ConAgra Foods, n.d.). ConAgra made no mention of its efforts to track and locate the source of the bacteria outbreak. In matters of food borne illness outbreaks, stakeholders need to know the efforts being made to rectify the situation. They especially need to know what to look out for, which products are potentially contaminated, and what they can do to protect themselves. While this may seem counterintuitive to organizations, they must communicate both what is known and unknown. It is wrong to over reassure stakeholders with absolute statements, because at any given moment, things can change. Even worse, organizations may have to contradict themselves when new information emerges.

Present Risk Messages with Honesty

Risk and crisis messages are complicated because there are no absolutes. A risk message should carry with it some level of equivocality when the facts surrounding the crisis are still uncertain. For instance, overly-reassuring messages undermine trust and compromise credibility. On October 11, 2007, just 2 days after releasing a news article advising consumers not to eat their Banquet turkey and chicken pot pies, ConAgra Foods issued a voluntary recall of all varieties of Banquet brand frozen pot pies and all store brand frozen pot pies sold under the names of Albertson's, Hill Country Fare, Food Lion, Great Value, Kirkwood, Kroger, Meier, and Western Family (ConAgra Foods, n.d.). ConAgra Foods stated that to simplify the recall it included all brand varieties, including poultry and beef. When initially informed by health officials of the pot pie contamination, ConAgra Foods informed consumers specifically not to eat the turkey and chicken pot pies. Two days later, the company contradicted its initial message by issuing a recall that included the beef pot pies which, it stated, was to lessen confusion (ConAgra Foods, n.d.). As a result of this contradiction, ConAgra Foods' credibility suffered. ConAgra would have been better off telling consumers it was unsure about the outbreak and was doing everything it could to resolve the issue.

Meet the Risk Perception Needs by Remaining Open and Accessible to the Public

Accurately delivering risk messages can pose a challenge. An organization cannot expect to reach every stakeholder using only one or two information sources. We suggest that accessibility has many dimensions, including information form, receiver characteristics, location, and communication channels. Not addressing these dimensions limits the effectiveness of the message. In its first news release, ConAgra Foods provided an address for consumers to use when returning the contaminated product labels and a telephone number for consumers with questions about the pot pies. Their web site was also mentioned as a source of information. The first news release included a statement outlining its effort to improve the product cooking instruction labels. In the two news releases posted on the ConAgra web site immediately following the recall, no mention was made of the Company's intention to communicate with the public. Besides what was broadcast on the news and in the initial press releases, ConAgra Food's only attempt to communicate with consumers about their concerns was made in a third frequently asked questions (FAQs) news release on its web site. This release addressed 12 consumer FAQs concerning the Banquet brand pot pie recall (ConAgra Foods, n.d.). In these 12 FAQs, ConAgra addressed all of the issues would have appropriate in the company's initial response. For instance, ConAgra stated it was working with the USDA to investigate the issue, discussed *Salmonella* and its associated risks, and clarified where consumers or someone they knew had been sick could call (ConAgra Foods, n.d.).

Communication about foodborne illness cases needs to be as thorough as possible. Relying on the Internet as the main source of information is ineffective because, as is discussed above, not everyone has access to or prefers to use the Internet as a primary information source. Any new information about the foodborne illness, including efforts to identify and correct the problem, should be communicated often to stakeholders.

Collaborate and Coordinate About Risk with Credible Information Sources

In an attempt to retain their reputation, organizations strategically attempt to shift the blame (Benoit, 1995). ConAgra Foods initially blamed customers for the outbreak

(ConAgra Foods, n.d.). It blamed customers in two ways: first, they were not reading the directions, and second, they were not cooking the product properly. ConAgra Foods stated it would be working with the USDA to clarify the cooking instructions. The company emphasized that it was "already revising its packaging to more clearly illustrate different cooking times for Banquet pot pies related to varying wattages of microwaves" (ConAgra Foods, n.d.). After the recall, ConAgra Foods added a new microwave cooking section to its Web site, but there were only a few sentences on bacteria, and three links: microwave oven basics; handling, storing and preparing frozen microwavable foods; and preparing popcorn in a microwave. The web page directed the reader to the Partnership for Food Safety Education (PFSE) web site for details on eliminating bacteria. There were also references to a government food safety web site and the FSIS web site www.foodsafety.gov. In an effort to collaborate with other sources, ConAgra Foods partnered with Fight Bac, a food safety education program, and borrowed their "core four practices," which include clean, separate, cook, and chill (ConAgra Foods, n.d.). ConAgra Foods took immediate action to make an argument for who they thought was responsible and why. What the company did not do was clarify any questions about the cause of the outbreak or about which product it found to be the specific culprit. In light of this case analysis, three practical implications emerged.

Implications for Effective Risk Communication

Avoid Unethically Shifting the Blame in Risk Communication

First, organizations should not attempt to shift the blame to groups negative}ly affected by a crisis. In this case, without sufficient information, ConAgra shifted blame for the *Salmonella* outbreak to consumers who, it alleged, were not cooking the product sufficiently. This is an inappropriate strategy and one that suggests that ConAgra was defensive in its post crisis communication, focused on protecting its image rather than protecting its customers' welfare. Organizations should avoid shifting blame until they have substantive evidence that in fact someone else is indeed fully responsible.

Avoid Over-reassuring in Risk Communication

Second, it is best not to make absolute statements about a crisis until all of the facts have been gathered. An overly-assuring statement made early on during a crisis can create two problems. First, it can make the public think something is being covered up, because most people understand there is an innate uncertainty in crisis situations. ConAgra spoke too soon and with too much certainty about which pot pies were contaminated and which were not. Because its statements were initially incorrect, ConAgra actually increased the uncertainty and created more confusion.

ConAgra should have maintained more flexibility until sufficient information was available.

Risk Communication should be Culturally Sensitive

Third, when faced with a crisis it is best to communicate clearly and make information pertaining to any unfamiliar conditions accessible. This becomes particularly important in the case of foodborne illnesses. Not all stakeholders can understand the language in risk warnings. If, for example, people are warned not to eat a product because it could have *Salmonella* contamination, one cannot assume everyone knows what *Salmonella* is and/or what symptoms are associated with a *Salmonella* infection. Initial risk messages must be thorough enough to inform every possible stakeholder. This may mean making specific warnings, with descriptive statements about the infection, in many languages. Risk communicators must account for varying risk tolerances along with various understanding of the potential risk in their messages. Finally, these messages should not be limited to the company web site; rather, they need to be widely accessible to many different audiences. By casting a wide net, an organization can work to create understanding among stakeholders and lessen the overall impact of uncertainty as well.

Foodborne illness in the U.S. is increasing in prevalence. Each case provides a challenge for the organization responsible. These cases continue to illustrate both effective and ineffective ways to respond to and communicate risk. In the face of a crisis, stakeholders need someone to turn to for answers. An organization's response to a crisis reveals what its values. If an organization appears more concerned with its bottom line than with its stakeholders' safety, the consequences can be devastating to the organization's reputation. Consequently, preparing in advance to respond appropriately is worth the effort.

References

Associated Press. (2007a, October 10). ConAgra shuts down pot pie plant over *Salmonella* link. Retrieved January 17, 2008, from www.FoxNews.com.

Associated Press. (2007b, October 10). Warning of a possible salmonella-pot pie link. *The New York Times*, p. A.21. Retrieved February 14, 2008, from ProQuest, http://proquest.umi.com.

Benoit, W. L. (1995). *Accounts, excuses, and apologies: A theory of image restoration strategies*. Albany, NY: State University of New York Press.

Bratton, A. J. (2007, October 12). Maker recalls all banquet and store brand pot pies. *South Florida Sun-Sentinel*, p. A.8. Retrieved February 14, 2008, from ProQuest, http://proquest.umi.com.

Business briefs [City Edition]. (2007, October 12). *Lincoln Journal Star*, p. 13. Retrieved February 14, 2008, from ProQuest, http://proquest.umi.com.

CIDRAP News. (2007, October 12). ConAgra recalls pot pies as *Salmonella* cases rise. Center for Infectious Disease Research & Policy. Academic Health Center—University of Minnesota. Minneapolis, MN: Regents of the University of MN. Retrieved January 17, 2008, from www.google.com.

ConAgra Banquet Pot Pie Samonella victims now number 272, CDC says. (2007, October 31). *NewsInferno.com*. Retrieved April 5, 2008, from http://www.newsinferno.com/archives/1981.

ConAgra Foods. (n.d.). ConAgra Foods Company Web site. Retrieved from http://www.conagra-foods.com.

Eamich, A. (2007, October 11). Missouri firm recalls frozen pot pie products for possible *Salmonella* contamination. United States Department of Agriculture News Release. Retrieved January 17, 2008, from www.google.com.

Funk, J. (2007, October 15). Pot pie recall delay faulted ConAgra said to face liability. *Pittsburgh Post-Gazette*, p. A.1. Retrieved February 14, 2008, from ProQuest, http://proquest.umi.com.

Giant Food alerts customers to product recall by ConAgra Foods. (2007, October 12). *U.S. Newswire*. Retrieved February 14, 2008, from ProQuest, http://proquest.umi.com.

National Center for Food Protection and Defense. (2007, October 20). ConAgra Potpie Analysis. Retrieved November 10, 2007, from http://www.ncfpd.umn.edu/.

Reynolds, B. (2002). *Crisis and emergency risk communication*. Atlanta, GA: Centers for Disease Control and Prevention.

Rosetta, L. (2007, October 25). Despite tainted food-related recall, pot pies stayed on some Davis County store shelves. *The Salt Lake Tribune*. Retrieved Febreuary 14, 2008, from ProQuest, http://proquest.umi.com.

Ruff, J. (2007a, October 11). Financial hit likely limited for ConAgra analysts note that only an advisory, not a recall, was issued over frozen pot pie concerns. *Omaha World-Herald*, p. D.1. Retrieved February 14, 2008, from ProQuest, http://proquest.umi.com.

Ruff, J. (2007b, November 15). ConAgra's menu again has pot pies: After a recall linked to salmonella illnesses, the chief executive apologizes and says the products will start shipping soon. *Omaha World-Herald*, p. D.1. Retrieved February 14, 2008, from ProQuest, http://proquest.umi.com.

Ruff, J. (2007c, December 21). Second-quarter earnings ConAgra profit increases 14.9 percent; the rise comes despite a $27 million expense because of pot pie recall in October. *Omaha World-Herald,* p. D.1. Retrieved February 14, 2008, from ProQuest, http://proquest.umi.com.

Schneider, C. (2007, October 10). Salmonella linked to ConAgra pot pies: Two Georgia cases—Consumers advised to check frozen chicken or turkey pot pie products. *The Atlanta Journal-Constitution*, p. B.5. Retrieved February 14, 2008, from ProQuest, http://proquest.umi.com.

Ulmer, R. R., Sellnow, T. L., & Seeger, M. W. (2007). *Effective crisis communication: Moving from crisis to opportunity*. Thousand Oaks, CA: Sage.

Weise, E., & Schmit, J. (2007, October 12). ConAgra recalls all frozen pot pies; product has been linked to salmonella outbreak. *USA Today*, p. B.1. Retrieved February 14, 2008, from ProQuest, http://proquest.umi.com.

William Marler, food safety attorney, urges ConAgra to recall all Banquet pot pies immediately to protect the public. (2007, October 10). *Business Wire*. Retrieved February 14, 2008, from ProQuest, http://proquest.umi.com.

Part III
Applications of a Message-Centered Approach to Risk Communication

Chapter 10
Toward a Practice of Mindfulness

There's more than a grain of truth to the saying that when all you have is a hammer, every problem looks like a nail.
–(Weick & Sutcliffe, 2007, p. 90)

Bazerman and Watkins (2004) characterize risk management with the apparent oxymoron of "predictable surprises" (p. 1). In other words, risk situations are fraught with potential harm, yet the inherent uncertainty of risk means we can never know when or if a problem will erupt. A *mindful* approach to risk management accounts for both the looming threat and intrinsic uncertainty present in all risk situations. By contrast, the "rigid reliance on old categories" results in routine or mindless reactions that are insensitive to emerging risks (Langer, 1989a, p. 63). Pragmatically, a mindful approach requires that risk managers sustain a "high level of sensitivity to errors, unexpected events, and–more generally–to subtle cues suggested by the organization's environment or its own processes" (Levinthal & Rerup, 2006, p. 503). This sensitivity to subtle changes or early warning signs ultimately enables organizations to address emerging risks before they progress into crises.

Obviously, organizations and agencies cannot monitor every potential element of risk every minute of every day. To do so is impossible and the attempt to do so depletes essential resources and confounds the decision-making process (Fiol & O'Connor, 2003). On the contrary, Bazerman and Watkins (2004) explain that a mindful approach to risk management allows organizations to establish adequate "triggering and response procedures" (p. 172) as they monitor risk. Such procedures "establish appropriate thresholds, or rules that determine when changes in key measures trigger action" (p. 172). Without such triggering procedures, organizations risk "either under-responding to significant changes or over-responding to statistically insignificant fluctuations" (p. 172).

In this chapter, we first provide an overview of the mindfulness concept. We offer particular emphasis on the role of risk communication in establishing an organizational culture of mindfulness. Next, we introduce the practical strategies for high reliability organizations as a generalizable means of achieving mindfulness in risk management. We then discuss the importance of mindfulness for achieving the level of convergence needed to effectively monitor and communicate about risk. In addition, we explain the relevance of organizational learning to mindfulness. We conclude with several complicating factors that pose particular challenges to achieving a mindful level of convergence.

T. L. Sellnow et al. *Effective Risk Communication*

DOI: 10.1007/978-0-387-79727-4_10

Mindfulness as Questioning the Routine Response

The concept of mindfulness, which originated in Buddhist philosophy, denotes "attentiveness to the present, rather than the faculty of memory regarding the past" (Thomas, 2006, p. 277). Ellen Langer (1989a) applied the mindfulness concept to research in psychology. Mindfulness has since been applied to the study of education, medicine, organizations, and communication. Langer (1997) explains that, regardless of the activity, a mindful approach inherently includes the following three characteristics:

- The continuous creation of new categories
- Openness to new information
- An implicit awareness of more than one perspective (p. 4)

Conversely, mindlessness invariably involves:

- An entrapment of old categories
- Automatic behavior that precludes attending to new signals
- Action that operates from a single perspective (p. 4)

Langer and Moldoveanu (2000) identify several advantages of mindfulness clearly relevant to risk management:

- Greater sensitivity to one's environment
- More openness to new information
- Creation of new categories for structuring perception
- Enhanced awareness of multiple perspectives in problem solving. (p. 2)

Langer and Piper (1987) add: "Failing to respond mindfully to ones environment results in 'unnecessary debilitation' " (para. 2).

For those who wish to instill in their employees a mindful approach, the challenge is to identify the appropriate types information to monitor. Langer recognizes the potential communication overload that can arise from attempts to monitor everything in one's environment. Langer and Piper (1987) explicate this dilemma: "The problem is that one cannot mindfully process every piece of information and what is irrelevant today may be relevant tomorrow" (para. 7). Langer (1989b) sees the ability to focus on context as the means for avoiding such mental exhaustion:

Whether one is responding mindlessly because of reliance on a structure built up over time or during a single exposure, when dealing with the world or any of its parts mindlessly, one is acting within a single context without active awareness of alternative conceptions. *Mindfulness* is essentially sensitivity to or awareness of contexts. Without this awareness, one cannot manipulate the context in which one finds oneself. One cannot improve performance, judgment, self-esteem, or health (p. 159). Ultimately, Langer suggests that "with adequate awareness of context, it may be possible to extend human performance far beyond currently accepted limits" (p. 168).

The quest for mindfulness poses a dilemma for organizations. Routine procedures are developed for purposes of efficiency and worker interchangeability. For example, french fries at a Little Rock McDonald's restaurant are prepared the same way at a Lexington McDonald's. For many organizations, such consistency is the means to profitability. Furthermore, Langer (1989b) explains, routines like McDonald's were, at some point, created by individuals in a highly mindful state. Considerations of food safety, worker safety, as well as efficiency were likely all included in the development of this routine. Maintaining this mindful state while enacting such routines requires individuals to engage in "creating (noticing) multiple perspectives, or being aware of context" (p. 138). For Langer (1989b), the trouble occurs when these routines become insensitive to environmental changes. If, for example a routine is "rule-governed rather than rule-guided" individuals become "mindlessly trapped by categories that were previously created when in a mindful mode" (p. 139).

Levinthal and Rerup (2006) argue that routines can be enacted in a way that sustains mindfulness. Acknowledging that workers have a "finite capacity for mindfulness," Levinthal and Rerup contend that organizations can maintain mindfulness through the "*recombination* of existing routines" (pp. 505–506). They liken this form of recombination to English Grammar. Elements of a routine can be recombined in the same way that a group of words can produce a variety of sentences. For example, some nuclear power plants alter the format of paperwork for safety inspectors to help them avoid falling into a mindless routine. The inspection remains a routine procedure; however, the reporting process is continually altered. Through recombination, existing knowledge, such as that accumulated from routine behavior, "is used as building blocks" for improvised or innovative behavior (p. 506). In the end, such recombination seeks to develop a mindful culture that is "concerned with how adequately people can convert experience into reconfigurations of assumptions, frameworks and actions, as well as how they legitimate learning from near misses and close calls" (p. 506).

Routine procedures that give way to the enforcement of rules rather than sensitivity to context can also diminish an organization's capacity for functioning ethically. Thomas et al. (2004) explain that "the once fashionable notion that business ethics could be safely relegated toward the bottom of the corporate 'to do list' exists no longer" (p. 56). Any remnant of such thinking dissipated with the recent shameful demise of such corporate powerhouses as Enron, WorldCom, and Tyco. To reduce the risk of similar crises, Thomas et al. advise organizations to instill a culture of ethics mindfulness. They define ethics mindfulness as "a form of self-regulation that causes one to behave with an ethical consciousness from one decision or behavioral event to another" (p. 61). Thomas et al. contend that ethics mindfulness cannot begin in an organization until "the leaders develop and display personal ethics mindfulness, thus becoming models for positive learning by others" (p. 61). In Chapter. 11, we discuss the importance of ethics in risk communication much further. Ideally, a culture of ethics mindfulness promotes a system of priorities and rewards where "employees are more concerned with integrity of the workplace than with rules and sanctions" (p. 64).

The importance of mindfulness in risk management is clear. If employees are not mindful, they cannot effectively observe and report subtle warning signs that a crisis may be emerging. Yet, incorporating mindfulness without overwhelming the attention capacity of employees is a delicate matter. Weick and Sutcliffe (2007) offer the consistent strategies of high reliability organizations as a model for achieving the appropriate degree of mindfulness.

High Reliability Organizations as Models of Mindfulness

Weick and Sutcliffe (2007) propose a pragmatic set of strategies designed to eliminate the *blind spots* that occur due to mindless routines. With regard to risk, blind spots "sometimes take the form of belated recognition of unexpected, threatening events" (p. 23). This failure to recognize emerging or expanding risks can result in a full-blown crisis. Weick and Sutcliffe purport that organizations and agencies can overcome blind spots through a mindful approach. They see mindfulness as an "enriched awareness" that "uncovers early signs that expectations are inadequate, that unexpected events are unfolding, and that recovery needs to be implemented" (p. 23). Wieck and Sutcliffe offer five principles, divided between anticipation and containment, for establishing mindfulness. We summarize each principle in the following paragraphs.

Anticipation

Weick and Sutcliffe (2007) describe anticipation as "preparing for the unexpected" (p. 45). As we have discussed throughout this book, risk is complex and indeterminate. Despite this complexity and uncertainty, warning signs are typically evident. The goal, then, is to anticipate problems based on early and subtle forewarnings. To effectively anticipate emerging or evolving risks, organizations must pay "mindful attention to three things: failure, simplification, and operations" (p. 45). They explicate a principle or strategy for each of these directives.

Preoccupation with Failure

A mindful approach to risk requires that failures, no matter how small, are noticed, evaluated, and corrected. Weick and Sutcliffe (2007) advocate treating "any lapse as a symptom that something may be wrong with the system, something that could have severe consequences if several separate small errors happen to coincide" (p. 9). In short, organizations must consider the "wider relevance" of small failures (p. 49). Successful organizations pride themselves on recognizing failures in the early stages, learning from them, and disseminating this knowledge throughout the organization.

Reluctance to Simplify

Although simplification in an organizational setting can enhance one's focus on the primary elements of a task, such simplification can also limit sensitivity to subtle warning signs. Weick and Sutcliffe (2007) warn that simplification fosters the perception of "superficial similarities between the present and the past" that can "mask deeper differences that could prove fatal" (p. 10). To avoid oversimplification, Weick and Sutcliffe recommend that organizations "simplify slowly, reluctantly, [and] mindfully" (p. 54). Doing so requires organizations to "welcome diverse experience, skepticism toward received wisdom, and negotiating tactics that reconcile differences of opinion without destroying the nuances that diverse people detect" (p. 10).

Sensitivity to Operations

Maintaining a mindful approach requires organizations to "be attentive to the front line, where the real work gets done" (p. 12). This is particularly true in the food processing and restaurant industries. Regardless of a product's quality or marketing, the organization is doomed to failure if the product is tainted in production or preparation. Weick and Sutcliffe (2007) explain that "sensitivity to operations is about the work itself, about seeing what we are *actually* doing regardless of what we were supposed to do based on intentions, designs, and plans" (p. 59). Weick and Sutcliffe identify three general threats to maintaining sensitivity to operations. First, organizations that place a high value on "quantitative, measurable, hard, objective, and formal" knowledge while disregarding "experiential knowledge" place themselves at greater risk. Weick and Sutcliffe explain that "doubt, discovery, and on-the-spot interpretation are hallmarks of sensitivity" (p. 60). A second threat is the tendency for routines to become mindless. To counter such mindlessness, workers should "rework the routine to fit changed conditions and to update the routine when there is new learning" (p. 61). Finally, organizations intensify their risk when they overestimate "their soundness" (p. 61). Such overestimations occur when organizations assume their invincibility after overcoming a near miss. The better reaction would be to consider what went wrong, how bad the situation could have been, and what changes are needed.

Containment

Containment is necessary when anticipation is unsuccessful and some degree of failure is manifested. Weick and Sutcliffe (2007) explain that the goal of containment is "to prevent unwanted outcomes *after* an unexpected event has occurred" (p. 65). As Bazerman and Watkins (2004) explain, the goal is always to predict and defuse risks. Yet, the uncertainty that is inherent in risk makes such prediction inconsistent.

Thus, organizations are wise to include containment as part of their risk and crisis planning. Weick and Sutcliffe offer two strategies for containment: "commitment to building resilience and deference to expertise" (p. 65).

Commitment to Resilience

A commitment to resilience recognizes the inability of any organization to account for all potential risks. The ultimate goal of resilience is to establish a plan for maintaining operations in the wake of an unforeseen and threatening event. For example, redundancy in operations, drills, crisis simulations, and the constant updating of crisis management plans are means of enhancing resilience. Because resilience is a response to a problem that has already occurred unexpectedly, the goal is always to correct mistakes "before they worsen and cause more serious harm" (p. 68). To maintain resilience, Weick and Sutcliffe (2007) propose that organizations must have three abilities. First, organizations must be able to "absorb strain and preserve functioning despite the presence of adversity" (p. 71). Second, the organization must have "the ability to bounce back from untoward events" (p. 71). Finally, resilient organizations have "an ability to learn and grow from previous episodes of resilient action" (p. 71).

Deference to Expertise

Weick and Sutcliffe explain that, as high reliability organizations seek to contain failure, they display deference to expertise when they "push decision making down and around" (Weick & Sutcliffe, 2007, p. 16). By doing so, "decisions are made on the front line, and authority migrates to the people with the most expertise, regardless of their rank" (p. 16). The wisdom of such deference is based on the assumption that those who work most closely with the procedures in question have the greatest knowledge. As a problem intensifies there is little chance that an individual atop an organization's hierarchy can make an accurate and appropriate decision from a distant office. As Weick and Sutcliffe observe, "Expertise is not necessarily matched with hierarchical position, organizations that live or die by their hierarchy are seldom in a position to know all they can about a problem" (p. 74).

The model of high reliability organizations offered by Weick and Sutcliffe (2007) offers practical strategies for adapting to emerging risks. This adaptation is achieved when organizational members limit cognitive distractions and work toward "the vividness of a better focused mind and wisdom" (Weick & Putnam, 2006). The wisdom acquired through mindfulness is based on the convergence of observations shared collaboratively by organizational members.

In the next section, we discuss the relationship between mindfulness and convergence at the individual and collective levels.

Mindfulness and Convergence

The concept of mindfulness is germane to risk communication on both the individual and collective levels. As organizational members engage in mindfulness, they observe the nuances in their routines, interpret them, and communicate about them. On an individual level, mindfulness requires that workers remain attentive to both their observations and their communication of these observations to others. On a collective level, members of an organization are asked to consider, debate, and act upon the inclusive knowledge base comprised of the mindful observations of both themselves and their peers. Consistent observations and interpretations produce the form of convergence that has the capacity to foster innovation. A failure to interact and communicate effectively leads to missed opportunities for the organization to adapt and fortify its management of an evolving risk.

Individual Mindfulness

On an individual level, workers must be mindful in both their observations and interpersonal communication. If individuals observe a subtle failure, the observation is of no value unless they can communicate that observation in an accurate and compelling manner. Individual mindfulness is predicated on an individuals' ability to notice subtle changes in their working environment. Burgoon et al. (2000) characterize such changes as "encountering an unfamiliar setting or routine, failing to bring about desired goals and subgoals, having completion of a planned course of action thwarted, or projecting that one's intended actions may have adverse effects" (p. 111). All such observation should "make interactants more mindful about their own and others' behavior" (p. 111). Failure to see such nuances leads to maintaining the ongoing routine, even though the routine fails to account for an emerging or evolving risk. Once such observations are made on an individual level, they must be shared with co-workers.

The interpersonal communication process among co-workers is important to heightening individual mindfulness. Sharing concerns and drawing each other's attention to unexpected outcomes "may intentionally or incidentally alter the level of mindfulness with which information is processed" (Burgoon et al., 2000, p. 107). Burgoon et al. argue that lacking "communication skills to monitor their actions and adapt their messages, without the breadth of repertoire that enables flexible, novel thought processes to translate into creative action, a more mindful state may not lead to more successful communication" (p. 121). In other words, organizations cannot expect their workers to consistently exhibit individual mindfulness unless they encourage workers to report their concerns, empower workers to address emergency situations, and give workers opportunities for input in the management of emerging concerns.

In addition to observing and expressing accounts of unanticipated events in the work setting, individuals must interpret and evaluate the messages they receive from others. Certainly, not all observations made by one's co-workers merit attention. Simply ignoring all stated concerns, however, results in mindlessly advocating a routine that may be fatally flawed. Butler and Gray (2006) note, "mindless acceptance of information or data gives rise to a perception of certainty that can create premature commitment to a solution" (p. 215). Such premature commitment can lead to the escalation of a minor failure or warning to a full-blown crisis situation. For this reason, we encourage active listening for workers at all levels. Individuals should consider the observations of others, compare these observations to their own surveillance, and make informed decisions. Most importantly, when individuals see convergence between the messages of others and their own observations, they are obligated to act in any way they can to reduce the emerging risk. The potential for individuals to change an organization's outlook on a risk situation is based on its collective mindfulness.

Collective Mindfulness

Weick and Roberts (2001) contend that organizations achieve collective mindfulness through heedful behavior. They describe heedful actions with adverbs such as "critically, consistently, purposefully, attentively, studiously, vigilantly, conscientiously, pertinaciously" (p. 263). In this environment, the organization's leadership seriously considers individuals' reports of unanticipated occurrences. The organization *makes sense* of its environment through the heedful interpretation of these messages. Seiling and Hinrichs (2005) see organizational leaders as "sensemanagers" whose primary responsibility is to engage organizational members in "mindfulness and constructive accountability" (p. 82). They depict organizing as a sensemaking process where "organizational members interpret their environment in and through interactions with others, constructing accounts that allow them to comprehend the word and act collectively" (p. 83). Through such shared accounts of what is taking place, organizational members collectively interpret events, set priorities, and enact response strategies.

In essence, organizations build a collective mindfulness through convergence. As accounts overlap in their description and interpretation, they foster a compelling case for change. Weick and Roberts (2001) explain: "Contributing, representing, and subordinating, actions that form a distinct pattern external to any given individual, become the medium through which collective mind is manifest" (p. 266). In this manner, an organization's collective mindfulness is constantly transforming. Achieving this state of heedful and perpetual change is difficult. Weick and Roberts argue that "it is easier for a system to lose mind than to gain it" (p. 279). If organizations become less receptive to conflicting messages and are unwilling to change, they become "heedless, careless, unmindful, thoughtless, unconcerned, indifferent" (p. 264).

To maintain collective mindfulness, experienced members must acquaint new members with the organization's commitment to an ongoing process of discovery.

Stories of near misses and past failures serve as warnings against complacency. By telling such stories, the experienced insiders often "resocialize" themselves, thereby heightening their own sensitivity (Weick & Roberts, 2001, p. 69). The prospects are grim for organizations that do not engage in this mindful transfer of knowledge. Weick and Roberts (2001) warn that "if insiders are taciturn, indifferent, preoccupied, available only in stylized performances, less than candid, or simply not available at all, newcomers are in danger of acting without heed because they have only banal conversations to internalize" (p. 269). Any such heedless behavior results in decay of the collective mindfulness.

Convergence is clearly present in both individual and collective mindfulness. Consistency and overlap in unanticipated outcomes and failures serve as convincing evidence for change in existing routines. As individuals and organizations engage in such mindful and heedful behavior, they generate evidence that can be used for organizational learning by other parts of the organization, by similar organizations, and even by entire industries. We discuss this form of learning in the next section.

Mindfulness and Organizational Learning

Engaging in either collective or individual mindfulness contributes to organizational knowledge. As such, mindfulness is a key part of the learning process. Langer (1997) proposes that "*sideways*" learning allows even novices to engage into a mindful state (p. 23). As opposed to top-down learning, discursive lecturing, bottom-up learning, or repeated practice of a new activity, sideways learning requires "learning a subject or skill with openness to novelty and actively noticing differences, contexts, and perspectives" (p. 23). When individuals or organizations engage in sideways learning, their behavior is guided "in the present, rather than run like a computer program" (p. 23). As individuals and organizations learn to redirect their activities in risk situations, staying in the present is essential. Doing so requires letting go of or forgetting routine behaviors that are no longer effective. Langer argues that this process of forgetting is a daunting task. She claims, "it is easier to learn something the first time than it is to unlearn it and learn it differently" (p. 85). Without such forgetting, our actions in the present are bound by mindless considerations of the past.

In contrast to Langer's (1997) sideways view of learning, Weick (1995) sees mindful learning as "*learning in reverse*" (p. 184). As individuals make decisions and plans, they do so prospectively. Even the most thorough planning cannot, however, completely account for the unforeseeable future: "Prospective decisiveness gets derailed over and over by unexpected events and unanticipated consequences of initial actions" (p. 184). The objective, then, is to interpret unanticipated consequences mindfully in a manner that enables the organization to establish a history that reveals opportunities for learning. By doing so, learning in reverse can actually enhance an organization's confidence for dealing with similar problems in the future. Weick explains that, ideally, such "decisive retrospect can facilitate subsequent decisiveness by creating expectation and scripts that function like self-fulfilling prophecies" (p. 184).

Sitkin (1996) agrees with Weick's notion that learning emerges from failure; going so far as to claim. "failure is an essential *prerequisite* for effective organizational learning and adaptation" (p. 541). The security afforded by continuous success can advance a lackadaisical or mindless perception of one's environment. As Sitkin explains, "The experience of failure produces a learning readiness that is difficult to produce without a felt need for corrective action" (p. 548). All failures, however, are not alike. Sitkin argues that the most intelligent or practical failures have the following five characteristics:

(1) they result from thoughtfully planned actions that
(2) have uncertain outcomes,
(3) are of modest scale,
(4) are executed and responded to with alacrity, and
(5) take place in domains that are familiar enough to permit effective learning (p. 554).

Without advance planning, the organization has little to review or correct based on failure. The failure must bring with it some level of uncertainty or surprise to generate new knowledge. If the failure is too severe, the organization must initially devote its energy to mitigation and management rather than learning. If organizations do not respond to the failure with alacrity, they risk losing out on valuable evidence that must be collected and considered immediately. Finally, failures offer the greatest opportunity if they occur within a domain over which the organization has some control and resources to manage. When these factors are present, the organization has a useful opportunity to learn and improve in a mindful manner.

Argyris (2001) explicates the communication strategies that enhance retrospective learning from failure. Most importantly, Argyris cautions against any reflective form of communication in the learning process that encourages workers to find fault with management and criticize policy. Doing so enables workers to shift from a mindful focus on their own behavior to an emphasis of past failings by management. Instead, Argyris advocates encouraging employees to "take active *responsibility* for their own behavior, develop and share first-rate information about their jobs, and make good use of genuine empowerment to shape lasting solutions to fundamental problems" (p. 88). By doing so, organizational members make a personal and ongoing commitment to maintaining a mindful approach to learning.

Learning opportunities are not restricted to the failures directly experienced by the organization. Rather, through mindful observation, organizations can learn from the mistakes of similar organizations. Huber (1996) explains that organizations experience vicarious learning when they "attempt to learn about the strategies, administrative practices, and especially technologies" of similar organizations (p. 135). When organizations experience crises, they frequently generate a wake of deliberative rhetoric for the relevant industry or industries to consider. For example, W. D. Stevens, president of Exxon Company, USA, created a retrospective analysis of myriad oil transportation policies in his response to the Valdez spill

(Johnson & Sellnow, 1995). Similarly, Jack-in-the-Box restaurants' *E. coli* outbreak caused the entire industry to reconsider its meat inspection and hamburger preparation routines. Organizations that fail to learn vicariously from their competitors' difficulties are likely to suffer even greater punishment or public rejection if they face similar crises (Sellnow & Brand, 2001).

Mindfulness can and should lead to an organizational learning process that expands an organization's overall intelligence on a long-term basis. The retrospective analysis of failures experienced and observed by organizations is essential to effective risk management. There are, however, circumstances that complicate the mindful learning process. We discuss several such circumstances in the final section of this chapter.

Complicating Factors to Mindfulness

Throughout this chapter, we have discussed the importance of mindfulness to maintaining effective risk communication on both the individual and collective levels. We have noted the lingering potential for organizations and their members to be lured into mindless routines. We see several other complications in an organization's environment that can lead to a less mindful approach. In this section, we discuss three such complications: hoaxes, bandwagons, and cross-cultural ignorance.

In Chapter 7, we provide a discussion of the New Zealand Foot and Mouth Disease hoax. The New Zealand case illustrates the way terrorists or other subversive groups can capitalize on the mindfulness of an organization. When subversive groups claim to have done harm, mindful organizations must at once enact all necessary response strategies while simultaneously communicating to the public that the threat is likely unsubstantiated. In such instances, mindfulness works against the organization's efficiency. Responding to false claims is not the best use of resources.

Burgoon et al. (2000) offer a positive response to the hoax complication. They argue that mindfulness provides a means for detecting deception, including hoax situations. Burgoon et al. feature selective mindfulness as a balance between identifying deception and remaining cautious. They explain that selective mindfulness "in terms of both frequency with which one becomes suspicious and the communication signals that receive close scrutiny, is more likely to lead to improved deception detection" (p. 115). This scrutiny can be established while "maintaining the flexible, tentative stance toward the veracity and validity of information that is the hallmark of mindfulness" (p. 115). In this manner, selective mindfulness offers a potential solution to the costly distraction of hoaxes.

The bandwagon effect can also produce a complicating factor for typically mindful organizations. "Bandwagons are diffusion processes whereby individuals or organizations adopt an idea, technique, technology, or product because of pressures caused by the number of organizations that have already adopted it" (Fiol & O'Connor, 2003, para. 2). The fear is that organizations will compromise their safety standards or mindfulness to risk in order to keep pace with competitors. The potential for mindless repetition increases as organizations continue mindlessly imitating the actions of their peers and competitors. For example, despite repeated warnings lending institution after lending institution continued to provide high-risk mortgages in the middle of this decade. The eventual result was tragic for the homeowners, debilitating for the lenders, and devastating to the national economy. A similar example involves the seemingly mindless shift to importing pet food products from China. When melamine was discovered as the cause of illness in pets, an investigation revealed that dozens of pet food products were involved due to an industry-wide shift toward importing ingredients from China. This bandwagon effect occurred despite the increased risk. Fiol and O'Connor argue that bandwagon behavior emerges when companies focus solely on success. They argue that organizations can maintain mindfulness through a seemingly paradoxical focus on both success and failure. In other words, as organizations visualize success, they must temper this vision with a mindful consideration of the worst-case scenario. This balanced approach discourages organizations from mindlessly imitating bandwagon behavior.

As we discussed in Chapter 3, cultural barriers can impede risk communication. Thus, we see cultural differences as another complicating factor for maintaining mindfulness. Thomas (2006) explains: "Mindfulness operates by establishing the opportunity to consider a range of behavioral options based on knowledge of how cultures vary and how culture affects behavior" (p. 86). Any mindless assumption that one risk message "fits all" listeners will likely result in discounting a notable portion of one's audience. For example, previous research completed by the National Center for Food Protection and Defense found that risk messages failed to reach as much as 20% of the population when standard communication channels such as network television and newspapers were used. Simply put, informational needs, preferences, and understanding vary among cultures. Thus, organizations must practice cultural sensitivity in both internal and external communication of risk messages. When responding to cultural needs, Thomas explains, mindfulness requires "choosing the appropriate behavior from a well-developed repertoire of behaviors that are correct for different intercultural situations and also extrapolating to generate new behavior" (p. 88). Achieving mindfulness of this sort allows organizations to see cultural differences as a potential resource for discovery rather than as a barrier or insignificant factor.

Complicating factors such as deception, bandwagon behavior, and cultural differences may appear daunting to those who desire to develop a mindful approach to risk communication. For each of these challenges, however, solutions are available. In the end, the added effort necessary to maintain a mindful approach is certainly a worthwhile investment.

Summary

The uncertainty inherent in risk requires that individuals and organizations remain constantly vigilant. Applied research in mindfulness offers practical suggestion for efficiently and effectively maintaining such watchfulness. Through the continuous creation of new categories, openness to new information, and an implicit awareness of multiple perspectives, organizations and individuals can observe, interpret, and evaluate subtle environmental changes that pose potential risks. The strategies used by high reliability organizations offer a practical means for introducing mindfulness to organizations of all types. Although organizing necessitates some degree of routine behavior, mindfulness allows for the ongoing assessment and innovation of habitual behavior. This assessment begins at the individual level and, through the recognition of convergence, expands to form a collective mindfulness for the entire organization. Mindfulness ultimately enhances an organization's knowledge base through organizational learning. This learning process is challenged by complexities such as deception, bandwagon tendencies, and cultural differences. These challenges can be overcome with a persistent dedication and balanced approach to mindfulness.

References

Argyris, C. (2001). Good communication that blocks learning. In *Harvard business review on organizational learning* (pp. 87–110). Cambridge, MA: Harvard Business School Publishing Corporation.

Bazerman, M. H., & Watkins, M. D. (2004). *Predictable surprises: The disasters you should have seen coming and how to prevent them.* Boston, MA: Harvard Business School Press.

Burgoon, J. K., Berger, C. R., & Waldron, V. R. (2000). Mindfulness and interpersonal communication. *Journal of Social Issues, 56*(1), 105–127.

Butler, B. S., & Gray, P. H. (2006). Reliability, mindfulness, and information systems. *MIS Quarterly, 30*(2), 211–224.

Fiol, C. M., & O'Connor, E. J. (2003). Waking up! Mindfulness in the face of bandwagons [Electronic version]. *Academy of Management Review, 28*(1), 54–70.

Huber, G. P. (1996). Organizational learning: The contributing processes and the literatures. In M. D. Cohen, & L. S. Sproull (Eds.), *Organizational learning* (pp. 124–162). Thousand Oaks, CA: Sage.

Johnson, D., & Sellnow, T. (1995). Deliberative rhetoric as a step in organizational crisis management: Exxon as a case study. *Communication Reports, 8*(1), 53–60.

Langer, E. J. (1989a). *Mindfulness.* Reading, MA: Addison-Wesley Publishing Company, Inc.

Langer, E. J. (1989b). Minding matters: The consequences of mindlessness-mindfulness. In L. Berkowitz (Ed.), *Advances in experimental psychology* (pp. 137–173). San Diego, CA: Academic Press, Inc.

Langer, E. J. (1997). *The power of mindful learning.* Reading, MA: Addison-Wesley Publishing Company, Inc.

Langer, E. J., & Moldoveanu, M. (2000). The contrast of mindfulness. *Journal of Social Issues, 56*(1), 1–9.

Langer, E. J., & Piper, A. I. (1987). Prevention of mindlessness [Electronic version]. *Journal of Personal and Social Psychology, 53*(2), 280–287.

Levinthal, D., & Rerup, C. (2006). Crossing an apparent chasm: Bridging mindful and less-mindful perspectives on organizational learning. *Organization Science, 17*, 502–513.

Seiling, J., & Hinrichs, G. (2005). Mindfulness and constructive accountability as critical elements of effective sensemaking: A new imperative for leaders as sensemanagers. *Organizational Development Journal, 23*(3), 82–88.

Sellnow, T. L., & Brand, J. D. (2001). Establishing the structure of reality for an industry: Model and anti-model arguments as advocacy in Nike's crisis communication. *Journal of Applied Communication Research, 29*, 278–295.

Sitkin, S. B. (1996). Learning through failure: The strategy of small losses. In M. D. Cohen, & L. S. Sproull (Eds.), *Organizational learning* (pp. 541–578). Thousand Oaks, CA: Sage.

Thomas, D. C. (2006). Domain and development of cultural intelligence: The importance of mindfulness. *Group & Organization Management, 31*(1), 78–99.

Thomas, T., Schermerhorn, J. R., Jr., & Dienhart, J. W. (2004). Strategic leadership of ethical behavior in business. *Academy of Management Executive, 18*(2), 56–66.

Weick, K. E. (1995). *Sensemaking in organizations*. Thousand Oaks, CA: Sage Publications.

Weick, K. E., & Putnam, T. (2006). Organizing for mindfulness: Eastern wisdom and western knowledge. *Journal of Management Inquiry, 15*(3), 275–287.

Weick, K. E., & Roberts, K. H. (2001). Collective mind in organizations: Heedful interrelating on flight decks. In K. E. Weick (Ed.), *Making sense in the organization* (pp. 259–283). Oxford, UK: Blackwell Publishers Ltd.

Weick, K. E., & Sutcliffe, K. M. (2007). *Managing the unexpected: Resilient performance in an age of uncertainty* (2nd ed.). San Francisco: Jon Wiley & Sons.

Chapter 11
Ethical Considerations in Risk Communication

We do not act rightly because we have virtue or excellence, but we rather have those because we have acted rightly.
–Aristotle

Ethical issues emerge whenever a decision or action has the potential to affect another person. Therefore, when we make decisions about communication–what to communicate to whom and when and how to communicate it–we are inevitably making ethical choices (Johanneson, 1996; Seeger, 1997). Choices related to the communication of risks almost always have important ethical implications (Cohen, 1996; Johnson, 1999). The dissemination of information about environmental pollutants, risks of using tobacco, disease risks associated with life-styles, location of some risk-related manufacturing, storage, or disposal facilities all have the potential to significantly affect people. The decisions associated with communicating information related to food marketing, production, and food borne risks are also highly significant, given the potential impact on large groups of people (Singer & Mason, 2006). In deciding what to communicate, producers and processors may choose to withhold some information about a food product because the information is sensitive. Food marketers may try to spin nutrition messages. Advertisements may promote unhealthy eating habits or omit important nutritional information. Companies may be slow to issue contaminated product warnings or recalls, thus increasing the public's health risk.

As noted earlier, the Centers for Disease Control and Prevention (CDC) estimates that each year 76 million Americans get sick, more than 300,000 are hospitalized, and 5,000 die from food borne illness. With some 250 pathogens identified by the CDC as causing food illness, food safety is essential to public health protection. Agriculture and the food industry are massive multi-billion dollar economic and social forces with far-reaching impacts, profound political influence, and great potential to foster either good outcomes or to create widespread harm. Changes in agricultural practices can wipe out entire local economies and devastate fragile ecosystems. Large food retailers have the ability to alter food production practices simply by changing the way supplies are purchased, marketed, and distributed. On the other hand, changes in agricultural technologies can make food more nutritious, cheaper, and more widely available. Food is a uniquely personal product. Eaten, it becomes part of the person, as the old folk wisdom has it: "You are what you eat." Food is necessary for life, health, and well-being; it is not a luxury people can choose

T. L. Sellnow et al. *Effective Risk Communication*

DOI: 10.1007/978-0-387-79727-4_11

to do without. Moreover, eating habits are closely correlated with a wide array of diseases, including diabetes, heart disease, and many forms of cancer. Food also has important spiritual, cultural, and religious significance. For many, questions of food purity and wholesomeness are critical issues. For others, the questions are largely about food adequacy and simply having access to enough nutritious food. Feeding the poor is a tenet of many religions. For these reasons, food-related communication issues are inherently ethical issues.

In this chapter, we review ethical issues as they apply to communication about risks in general and about the risks associated with food. This includes messages created and disseminated by food producers and distributors and messages created and disseminated by regulatory agencies. We first describe ethics' functions in risk communication, including the role values play in ethical judgments and the ethical issues that must be considered at various points in the risk communication process. We also explore how ethics function in risk communication about food and food safety issues. This chapter then describes four basic ethics issues related to food: stakeholder views, significant choice and access to information, multiple constructions of risk, and issues of the legal versus value based strategies.

Ethical Decision-Making in Risk Communication

Ethics issues and ethical decision-making are inherent to being human. As humans, we naturally develop, learn, and acquire values. Values are the conceptualizations of what is good/bad, desirable/undesirable, appropriate/inappropriate, and right/wrong. These values are then used to inform decision-making and to critique our own and other's decisions and behaviors. We constantly make judgments about the ethics of telling the truth, withholding information, distorting the facts, or telling small "white lies." Judgments are regularly made about what constitutes fair and equitable treatment of others. Responsibility is another value that is frequently used to critique behaviors and inform decisions. Issues of social justice have become increasingly important standards for assessing the appropriateness of decisions, policies, and actions.

One of the most interesting and poignant examples of ethical choices and risk communication occurred in the long-running debate over smoking and health. When questions about the addictive nature of nicotine and the role of smoking in cancer and heart disease first began to emerge in the 1950s, the multi-billion dollar tobacco industry set out to counter these concerns. The industry embarked on a sophisticated public relations and issue management campaign designed to spread disinformation and confusion. The facts about health risks were withheld from the pubic or countered with industry-sponsored studies that discounted the link between smoking and disease. Even though the industry knew that nicotine was addictive, company leaders consistently claimed it was not. Some cigarette companies even advertised their products as healthy and featured doctors in their ads. The industry used its considerable economic and political power to overwhelm any messages that its products could cause harm. It was not until the Federal Cigarette Labeling and Advertising Act of 1965 that any product warning was mandated for cigarette packages. The Act required that the warning, "Caution: Cigarette Smoking May Be Hazardous to Your Health," be placed in small print on one side of each cigarette package. Even this warning was stated in very equivocal terms. More specific and detailed warnings were not mandated until the 1980s. For decades, people were denied accurate information about the risks of smoking.

Access

This example illustrates an important point about ethics and risk communication. Individuals cannot make informed choices about engaging in some behavior, or taking some risk, without the benefit of accurate information about that risk. In these cases, the lack of information interferes with the ability of the individual to make an informed personal choice. Withholding information, distorting the facts, seeking to create confusion, or lying about a product risk, such as was done with tobacco, means that consumers cannot exercise their fundamental human right to make an informed choice. When information about contamination of water supplies is withheld, members of the community are unable to make an informed choice about drinking the water. Consumers' ability to make decisions and judgments about what food is healthy or good for them depends on their access to accurate and understandable information. If nutritional information is withheld, the ability of consumers to make informed choices about the risks and benefits of certain foods is significantly reduced.

In Chapter 1, the concept of meaningful access was described. For access to be meaningful, pertinent information must be presented in a form the intended audience can understand, along with opportunities for audience members and members of the public to interact with key decision-makers. A common practice, for example, is to print important risk information in very small type, as was done with early cigarette warnings. While this practice saves space and is convenient for the manufacturer of a product, it may limit the ability of consumers to access the information. Similarly, risk information is often presented in highly technical terms, using jargon the average lay-person cannot understand. In these cases, accessibility to the information is also severely limited. Without meaningful access, the ability to weight alternatives, consider different options, and make decisions is undermined.

Values

The values that inform our own decisions and our judgments of others come from a variety of sources. They are part of our social conventions, and they form the underlying assumptions of legal codes and many government policies. We learn values from family, friends, faith-based organizations, community, and increasingly from the media. Values of justice, freedom, democracy, and open access to information come from our governmental institutions and traditions. Most formal religions include specific and detailed lists of "oughts" and "shoulds." Religious traditions often include unequivocal requirements for the use of language and communication. The ninth commandment from the Old Testament, "Thou shall not bear false witness against thy neighbor," is often interpreted as a general call for honesty and truthfulness.

The values that inform our ethical judgments, however, do not function in clean, concise, and autonomous ways. Rather, they function within a larger context of others' values. Within the context of any particular decision there may be dozens of ethical standards, values, needs, loyalties, and ethical traditions, all competing for prominence. In making choices about what food to purchase at a supermarket, for example, shoppers may consider the relative merits of regular and organic produce. Regular produce may be cheaper, more attractive, offer more variety, and be more widely available. Organic produce may seem healthier, more natural, and more environmentally friendly. At some level, the relative value of each type of produce is weighed in making purchasing decisions.

Choices about communicating any form of risk almost always involve completing values and interests. Risk communicators, for example, often must weigh the consequences of creating unnecessary concern and anxiety against the values of fully informing the public. When making the choice to evacuate a community due to a hurricane warning, officials must always consider the inevitable economic disruption and potential for injury and death that comes from a large exodus. There are often potential economic ramifications, such as the possibility of lost sales or even lawsuits, that risk communicators must consider. Some individuals, such as children or non-native speakers, may have a more limited capacity than others to understand and process risk information. In cases such as these, questions of equal access to information and fairness come into play.

Values always function in dialectic tensions with other values. In order for a decision to be made, one set of values must take prominence over another, either intentionally or unintentionally. Sometimes decision-makers are simply unaware of the ethical issues associated with their decisions. A consumer may not understand that child labor is used to manufacturer products in a developing country. The relative importance and prominence of various value systems are often debated. In some cases, these debates become public and develop in such a way that they are associated with specific interest groups. Often, the debates are adversarial and polarizing. For example, the values for treating animals in humane ways have become associated with animal rights groups such as People for the Ethical Treatment of Animals (PETA). Some groups argue that animals should have the same rights as humans. These groups are often in direct conflict with groups who view livestock exclusively as resources and commodities to be used for economic gain. Some environmental groups have argued that the presence of chemical pollutants, such as mercury or lead, at any level in drinking water is unacceptable. Manufacturers and industry groups typically argue that there are safe, allowable levels of such pollutants and that eliminating them completely is simply not practical. Debates about appropriate standards often digress into uncompromising positions, with interest groups tenaciously clinging to their values and associated ethical standards and refusing to consider others' positions. The only way to break such an ethical impasse is to agree on some common value system that can inform the decision.

As noted in Chapter 1, these value positions may also overlap or converge, creating a middle ground that represents an area of agreement based on the convergence of arguments. In the case of allowable levels of pollutants in drinking water, both environmental and industry groups might agree that lower levels are desirable and work toward the common goal of reducing contamination. In the case of animal rights, it might be agreed that, even if viewed as commodities, animals can be treated humanely. These convergence points in agricultural ethics help explain programs such as the Certified Humane Raised and Handled (CHRH) program for meat, poultry, eggs, and dairy products. Value convergence is a primary way in which ethical issues can be resolved.

Occasionally, the values represented in a particular circumstance are evenly balanced, creating an ethical dilemma for the decision maker. With some issues involving food and food production, decision makers must balance the interests of various stakeholder groups. In many instances, the needs and values of stockholders, consumers, producers, distributors, and regulatory agencies must all be taken into consideration as decisions are made. Balancing competing values can create significant dilemmas for decision makers.

When generally accepted ethical codes or values are violated, the actions and the persons and organizations responsible come under scrutiny and criticism. There is usually a process of determining who was responsible for the unethical act and an effort to sort out cause and blame. Often there are efforts to diffuse responsibility or shift blame, particularly if multiple agencies and organizations are involved. In the case of food contamination, for example, growers, shippers, processors, and regulatory agencies might each point fingers, seeking to shift blame and responsibility to others. This is what happened in the 2007 ConAgra Banquet chicken pot pie case. The company blamed consumers, claiming they failed to cook Banquet's "Not-ready-to-eat product" at a temperature high enough to kill harmful bacteria. The responsibility for the outbreak, the company claimed, was with the consumer not the producer.

Accountability

The process of sorting out blame and responsibility ultimately leads to an effort to determine accountability. Accountability literally means being able to offer an accounting or explanation of what went wrong and why. Accountability can result in a loss of reputation or legitimacy. In cases of dangerous or defective products, loss of reputation results in consumers' distrust, additional regulations, and, ultimately, losses in sales. In the case of tobacco discussed earlier, the industry was finally held accountable through a series of lawsuits and legislative initiatives. The industry was forced to acknowledge its past deceptive practices, adopt new restrictions on its marketing, and reimburse states for the health case expenses related to smoking. Most importantly, the industry agreed to fund public education programs about the risk of smoking.

It is important to note that ethical issues are evident throughout the risk communication process. Johnson (1999) described a nine-stage model of the risk communication process and identified ethical issues and choices associated with each stage. He notes that the first stage, identifying the issue that requires risk communication, often involves little choice and thus does not usually create significant ethical challenges. The choice of whether to communicate about risk is, however, an important ethical question. The second stage, setting goals, also includes a number of ethical questions. For example, some risk communication campaigns are designed to enhance understanding so that members of the public can make more informed choices. Others are designed to engineer consent, divert attention from the risk, or create confusion. Stage three involves knowing the issue, which often requires the risk communicator to depend on subject matter experts for technical information. At this stage, differences may emerge in opinion, perspective, and values. Johnson's fourth stage involves identification of audiences. Most risk issues have the potential to affect a variety of audiences, such as community members, employees, regulatory agencies, and special interest groups. Ethics would require ensuring that all relevant parties are involved. Knowing the constraints of the situation, the fifth stage, is a practical as well as ethical consideration. Situational constrains may limit the audiences who may reached, the methods of communication that are available, and the timing of the message. Audience analysis, stage six, is an opportunity to include the various groups and communities that have a stake in the risk issue. At this point in the risk communication process, the message is developed. This is the stage where a majority of ethical questions and decisions become evident. These decisions include how much information to communicate, what topics to cover, how to present these topics in terms of frames and formats, and how the message is assessed (Johnson, 1999, p. 341). The eighth stage involves ethical choices about channels and media for delivering the message. Some channels and media, for example, many allow for two-way communication about risk and the development of a risk dialogue. Some media may be more accessible to the audiences affected by the risks. Finally, Johnson's model suggests that all risk communication programs be evaluated for both their effectiveness and adherence to standards for ethical conduct.

Risk communication is inherently an ethical domain and ethical questions and issues are evident throughout the process of planning, executing, and evaluating a risk communication program. The practice of risk communication involves sorting out and balancing competing and contradictory values associated with diverse audiences. Each domain of risk communication includes a unique set of demands, contingencies and values. This includes environmental risks, risks associated with life style, and risks associated with food and food safety.

Ethics in Food Issues

As noted earlier, food is an area associated with profound ethical questions and choices. Zwart (2000) notes that even the ancient Greeks discussed the ethics of food in developing their concept of dietetics. Individuals learn from parents, education systems, and public service announcements that some foods, if eaten in excess, can cause health problems. In this line of reasoning, immoderate consumption of these foods is considered bad, undesirable, and wrong. When making choices about ordering food from a local restaurant, these values inform our decisions. We would expect that most consumers would at least try to avoid making an entire meal of the most unhealthy, high caloric items. When we see people eating excessively, we call them gluttons and we judge this behavior negatively. In his *Moralia in Job*, Pope Gregory the Great (d. 604) cited gluttony as one of the seven deadly sins.

Peter Singer (2002), one of the founders of the contemporary applied ethics movement, has written extensively about food issues. His work, *Animal Liberation*, focused on the morality of animal exploitation and included the use of animals in agricultural practices. Many of the recent changes resulting in a more humane treatment of agricultural animals can be traced to Singer's work. His later works have a broader focus on individual food choices and their impact on animals, workers, indigenous cultures, the larger society, and the environment. Singer & Mason (2006) describe the profound ethical consequences when food choices drive markets and markets in turn drive agricultural practices. In some cases, market forces lead to unethical practices, as demands for lower prices create factory farming pressures resulting in the inhumane treatment of animals. In other cases, markets respond with ethical standards and practices. Ben and Jerry's ice cream company's recent decision to transition to certified humane eggs in its products is one example of this type of response. Like many other food companies, Ben and Jerry's features its humane certification in its marketing efforts. The basic choices consumers make about food, then, have important ethical dimensions and the information the consumer has access to impacts the ability to make these informed choices.

In the following section, we review four basic issues of ethics related to risk communication and explore how they function within the context of food. These include stakeholder views, significant choice and access to information, multiple constructions of risk, and issues of the legal versus value based strategies.

Stakeholder Values

One of the first steps in making ethical and responsible decisions is to identify the groups or stakeholders who have an interest in the decisions (Ulmer, 2001; Ulmer & Sellnow, 2000; Ulmer et al., 2007). Sometimes, decisions inadvertently offend some group because the decision makers simply failed to recognize or consider that groups' values during the decision-making process. Once stakeholders are identified, it is possible to establish a two-way exchange of information, ideas, and perspectives so that their views are not ignored in decision-making. This includes diverse information, ideas, and perspectives about risk. Ensuring that all stakeholders have a voice is consistent with larger values regarding democratic representation, equality, and participatory decision-making.

Stakeholder theory developed out of work in strategic planning. Strategic plans are founded on an understanding of the resources, needs, and interests of an organization. This requires identifying and analyzing those groups who have a stake in the organization. Stakeholder theory quickly evolved into a useful way of examining corporate social responsibility. Corporate social responsibility is generally defined as managing the organization and its operations so that there is an overall positive impact on society. This entails taking into account the needs and interest of all the various groups affected by the organization's actions.

Although stakeholders may involve a very broad set of groups, typically they are limited primarily to stockholders, employees, management, consumers, producers, distributors, suppliers, regulatory agencies, and the communities within which organizations do business. Increasingly, many ethicists suggest that organizations also have an ethical duty to the physical environment. Most producers of meat and poultry would agree that they have some duty to treat livestock humanely, and most farmers have a deep respect for the land. As food production has become global, stakeholder duties have extended to other cultures and communities.

Companies, including food companies, often find it difficult to understand and comply with cultural norms and values when they extend the practices into new regions. Values regarding food and business vary widely from culture to culture. Many recent issues of food safety, including *E. coli* outbreaks in green onions, hepatitis A in strawberries, and avian influenza (H5N7), have involved food produced under different regulatory systems in Asia or South America. These systems do not necessarily operate according to the same standards and values as companies in the United States.

As stakeholders expand and become more complex and diverse, it is increasingly difficult to balance their competing needs and values. Often, decisions makers cannot avoid privileging some stakeholders over others. Knowing what stakeholder values are evident in a situation will help decision makers act in an ethically responsible way. Phillips (2004) goes further and suggests that communication between an organization and its stakeholders is a moral obligation: "Individuals and groups who contribute to the organization should be permitted some say in how that organization operates" (n.p.).

As described earlier, stakeholders generally include the entire range of groups and individuals affected by and/or affecting an organization. In cases of agriculture and food production enterprises–which often have long multi-national supply networks, multiple components, and wide-spread plants making final pre-packaged products–stakeholders are numerous and diverse. For example, a food company may receive products from both domestic and international producers who compete with one another to produce the least expensive commodity. Food companies may have stockholders who primarily value return on investment and employees who primarily value wages, benefits, and good working conditions. The needs of these two stakeholder groups may be in conflict, resulting in competing arguments. When this happens, it may be necessary to balance needs or seek convergence between the various stakeholders.

The ethical obligation and communication practice called for by stakeholder theory involves understanding stakeholder needs, interests, and values and representing them in decisions. This includes the ways in which these stakeholders understand and frame risks. This is achieved through a dialectic exchange and ongoing two-way communication so that mutual understanding is achieved and tensions and conflicts, such as those discussed above, are clarified and considered, if not resolved. The result is a mutually beneficial relationship between the organization and its stakeholders and decisions that are ethically responsible.

Significant Choice and Right-to-Know

Earlier, we noted that consumers' ability to make informed choices about risks depends on the information to which they have access. In general, ethicists have emphasized that humans have a unique ability to make rational choices to weigh alternatives, consider arguments, and reach conclusions (Nilsen, 1974). This capacity for rational thought is enhanced when decision makers have free access to all available information and seriously undermined when information is withheld (Johnstone, 1981; Ulmer & Sellnow, 1997). There is an ethical obligation to provide all the relevant information whenever individuals face a *significant choice*.

The idea that information enhances the quality of decisions is well-entrenched in western thought and philosophy. Aristotle and many of the Greek philosophers believed that argument, as a way of considering all view points, led to the discovery of truth (Rowland & Womack, 1985). In fact, the Greeks specifically emphasized the same human capacity for rational decision-making in food choices (Zwart, 2000). This idea has been enshrined in democratic systems and institutions through the ideal of the "open market place of ideas." Within the public domain, the open market place, ideas and arguments compete with one another giving rise to the correct or most accurate ideas. In democratic systems, it is a free press that generally has the responsibility for providing open access to information so that citizens are informed in making "significant choices." Sunshine laws, open meeting acts, and state and federal Freedom of Information Acts (FOIA) all serve to facilitate the free flows of information to the public so informed decisions can be made.

Significant choice has been encoded in a number of legal frameworks and cultural practices. One of the most far-reaching of such practices concerns the right-to-know. Right-to-know emerged as a general principle associated with environmentalism and the right of communities to know what chemicals are being released into their local environments. The Environmental Protection Agency (EPA) insists that all Americans should have the right-to-know the chemicals to which they are exposed in their immediate environment. Right-to-Know laws inform the public about such exposures. The public can then make informed choices about the risks they face and about limiting their exposure. The same principle applies to fish consumption advisories based on levels of mercury contamination, warning labels on tobacco products, or information about exposure to sexually transmitted disease through unsafe sex practices. Without adequate information, the public is simply unable to make informed and rational decisions about significant choices.

This generalized right to have access to information has also been evident in labeling laws and country of origin initiatives for food. A series of federal legislative mandates going back to the origins of the Food and Drug Administration (FDA) in the early 1900s have required listing information such as contents, weight, and the presence of artificial flavoring and colorings. Later, requirements were extended to ingredients. Many of these requirements were explicitly made to facilitate consumer comparisons and choices. In the 1990s, a series of federal efforts strengthened the requirements even further to include mandatory nutrition labeling for most foods, standardized serving sizes, and uniform use of health claims. In commenting on the initiative, then FDA Commissioner David Kessler, M.D., noted that the new food label provided Americans with an opportunity to make more informed and healthier food choices.

Food labeling is required for prepared and processed foods such as breads and cereals; canned, pre-packaged and frozen foods; and snacks, desserts, and drinks. General nutritional labeling is voluntary for raw fruits, vegetables, and fish. Additional legislation, the Food Allergen Labeling and Consumer Protection Act of 2004, addressed the informational needs of consumers allergic to specific foods. Allergies to foods and food products such as gluten, nuts, and shell fish are estimated to affect 2% of the adult population and 5% of the children. The problem also seems to be growing. In some cases, these allergies are life-threatening, even with very small exposure levels. Cross-contamination can also create severe allergic reactions. For at-risk consumers, accurate information on food labels is critical to making informed choices about food selections.

Food labeling is also controversial in that it sometimes places a heavy burden on producers. Producers with limited resources may not be able to meet all the labeling requirements. In addition, some producers argue that labeling calls attention to food product elements or processes that have not been shown to pose any risks. Food irradiation and bio-engineered foods, for example, have been targeted by some groups for food labeling, while proponents of these technologies have suggested that doing so simply creates the perception of risks where none exists. Similar arguments have been made about country of origin food labeling. Country of origin information is necessary for consumers to make choices about supporting local or regional economies. Consumers may also wish to choose foods that have smaller carbon footprints. Many retailers post "locally grown" signs as a way of meeting this informational need. While a majority of the public believes country of origin labels would help consumers make informed food choices, producers fear the expense of tracking products through often very extended supply chains.

Significant choice and right-to-know are powerful ethical standards influencing much of risk communication (Johnson, 1999). They have broad applications to risk communication in a variety of contexts, including environmental risks, life-style risks, and issues of food and food safety. These standards have been encoded in a number of legislative initiatives, including federal food labeling regulations. Informing the public about the risks they face allows them to exercise their own perspectives, standards and values in making choices.

Multiple Constructions and Frames for Risk

We noted earlier that where there are ethics questions, diverse values are often in competition. In Chapter 1, we also discussed alternative frames of acceptance for risk and suggested that no single frame can be considered the *correct* or most accurate frame. These alternate frames include, among others, technical and scientific frames, cultural and social frames, and individual and personal frames. In matters of risk, the scientific and technical frame has traditionally been positioned as the only legitimate knowledge about risk, to the exclusion of lay persons and the general public from all risk discussions and decisions. Technical assessments of risk are often couched in scientific jargon and methods incomprehensible to lay persons. Moreover, technical knowledge is often centered in organizations, furthering the institutional perspective's prominence in risk assessment and decisions. Privileging the technical sphere violates the rights of all individuals affected by an issue, regardless of their background, education, or position. When the only legitimate argument is from science, the layperson has no voice in the decision, thereby no opportunity to gain entente. An ethical approach to risk communication, then, acknowledges that there are multiple frames for risk and that they are all legitimate. In this way, ethically-informed risk discussions are more inclusive of those who are actually at risk.

As we discussed in Chapter 3, community, cultural, individual, and personal constructions of risk are often anecdotal and based on specific experiences and individualized risk factors, while technical notions of risk are usually statistical and population-based. A particular community, however, may have direct experience with an environmental risk. The experience of a toxic spill contaminating water supplies and fisheries would naturally result in heightened concerns and sensitivity and different value systems about that risk factor. These conditions occurred in many Alaskan communities following the Exxon Valdez oil spill. Similarly, a community that has experienced a wide-spread foodborne illness outbreak may have a very different understanding of what constitutes an acceptable risk and may be much more risk averse than other communities.

Individual and personal constructions of risk function in similar ways. A person who has experienced a serious case of food borne illness, either directly or vicariously, is likely to have a different understanding of risk and a correspondingly lower risk tolerance. Research has also shown that risk perception varies according to demographic factors. Women, in general, are more accepting of risk messages while men are generally more willing to take risks. Younger people tend to be more risk tolerant (Mileti & Sorenson, 1990). Some demographic groups may not have the resources necessary to manage risks. People with lower incomes often do not have the resources to move away form the health risks created by environmental contaminants. In other cases, individuals with compromised immune systems or severe food allergies are often very risk averse. In these cases, they may have an increased need for risk information. Personal notions of risk are also influenced by family obligations. Small children, people with special needs, or older family members may significantly alter the way risk is framed.

In Chapter 1, we discussed Peter Sandman's conceptualization of risk as an assessment of hazard (or technical risk) plus outrage (social or cultural conceptualizations). Sandman's framework acknowledges the fact that multiple risk frames exist and should be acknowledged in any discussion of risk. The Food Ethics Council (2007) proposes inclusiveness, "involving and answerable to the people they affect" (n.p.), as a key food ethic. This inclusiveness is particularly relevant when considering multiple constructions and frames for risk.

To maintain an ethical stance in risk discussions and decision-making, multiple frames need to be both represented and taken into account. Those systems and structures that privilege only technical and scientific views, exclude lay persons and lay viewpoints from deliberations, and reject anecdotal accounts and personal concerns as "not scientific" fail to meet the test of inclusiveness. This is not to suggest that scientific and population-based risk assessments should not play a prominent role in public policy debate. Rather, it suggests there is an ethical obligation to acknowledge the legitimacy of other views about risk and to include them in a comprehensive risk dialogue. Doing so is both ethical and necessary for convergence.

Legal Versus Value Response

In addition to the tensions between various risk frames and perspectives, tensions also exist between traditional legal perspectives and larger value perspectives. These tensions are often manifest in open versus closed approaches to risk communication and associated efforts to manage short-term legal liability versus longer-term issues of organizational reputation and image. These tensions are most often felt when an organization has been accused of some wrongdoing and must respond.

Legal standards are usually grounded in larger social values and notions of ethical conduct, but concepts of what is legal and what is ethical diverge significantly. The law is usually drawn from minimalists' standards designed primarily to maintain basic social order and operation (Seeger & Hipfel, 2007). As described earlier in this chapter, food-labeling laws are based in a larger social value: consumers' right-to-know. Labels usually provide the minimal information necessary for consumers to make comparisons and informed choices. In practice, however, legal perspectives are generally directed toward limiting the liability that organizations may face when their conduct is assumed to create some harm. In these cases, the typical legal approach is to avoid revealing any information that may increase perceptions of wrong doing and enhance legal liability. This avoidance includes statements about concern and compassion that could increase the possibility of costly litigation (Martinelli & Briggs, 1998). Fitzpatrick & Rubin (1995) note the legal strategy for responding to accusations of wrongdoing usually includes saying as little as possible, using privacy issues as a justification, denying responsibility, and shifting blame where possible (p. 22).

In the case of the 1993 Washington State outbreak of *E. coli 0157* from undercooked hamburgers, for example, Jack-in-the-Box restaurants initially adopted what was primarily a legal strategy (Sellnow & Ulmer, 1995). This included denying responsibility and shifting blame. Later, the company changed its strategy to accommodate larger frameworks for ethical conduct. This included changing cooking temperatures to reduce the possibility of *E. coli* contamination and offering to pay the medical bills of those affected by the outbreak. In addition, the company was more open and apologetic for the harm.

Alternatively, in response to an October 1994 *Salmonella* outbreak associated with its ice cream, Schwan's Foods responded from what was clearly an ethical standpoint. The company quickly moved to recall the suspect product and offered to pay the medical expenses of those customers who became sick (Sellnow et al., 1998). The company used its fleet of door-to-door sales persons to track down the contaminated product and alert the public. Although the outbreak was later traced to poor sanitation practices by one of the company's trucking contractors, Schwan's made no effort to shift the blame or deny responsibility.

The contrast between the actions taken by Jack-in-the-Box and Schwan illustrates fundamentally different stances in response to accusations of wrongdoing. In one case, the company was slow to respond and initially sought to reduce its liability by denying responsibility and shifting blame. These and similar strategies would be described as traditional legal strategies in response to a risk. In the latter case, Schwan's moved quickly to correct the problem and accepted responsibility for the outbreak, exhibiting openness and concern for those individuals who had consumed the contaminated product. This approach privileges customer relationships, reputation, and long-term corporate responsibility image concerns in what might be termed a value-based response to risk. Some observers have suggested that a value-based response is more natural and honest than the kinds of strategic responses where consequences, including legal implications, are carefully weighed. In addition, these ethical stances may be associated with the long-term stability and survival of organizations.

While the legal response and the value response are not necessarily mutually exclusive, they are often positioned as competing ethical frames for communicating risk, where a legal response precludes a value response and vice versa. From the standpoint of risk communication ethics, the traditional legal standpoint has the potential of creating an additional level of criticism as the organization's response is judged as incomplete and failing to acknowledge responsibility. In contrast, a value based response grounded in right to know and accountability has the potential to bolster reputation and enhance positive stakeholder relationships.

Summary

Risk communication is itself a risky practice (Garland, 2003). The risk is increased when appropriate ethical standards are not applied. Ethics are part of any decision that has the potential to affect others and risk communication almost always includes that potential. In many cases, the ethics of risk communication are particularly profound as they may impact personal and community health, quality of life, livelihood and well being. Processes and practices of risk communication involve profound ethical questions in areas such as stakeholder views; significant choice; access to important technical, cultural, community information; personal risk constructions; and issues of the legal versus value based strategies.

At the very least, decision makers should be aware of the ethical issues inherent to the processes of risk communication. They should also seek stakeholders' input in the decisions that affect them and should make information available so when decisions are made, they are informed decisions. Risk communicators should work to understand and respect the diversity of stakeholder values and perspectives associated with a risk issue. Ultimately, the result is not only better decisions, but decisions that are more inclusive and ethical.

References

Cohen, V. (1996). Telling the public the facts–or the probably facts–about risks. In C. R. Cothern (Ed.), *Handbook for environmental risk decision-making: Values, perceptions and ethics* (pp. 103–114). New York: CRC Press.

Fitzpatrick, K. R., & Rubin, M. S. (1995). Public relations vs. legal strategies in organizational crisis decisions. *Public Relations Review, 21*, 3–21.

Food Ethics Council. (2007). Retrieved December 7, 2007, from http://www.foodethics-council.org/

Garland, D. (2003). The rise of risk. In R. V. Erickson & A. Doyle (Eds.). *Risk and morality*, 48–86. Toronto: University of Toronto Press.

Johnson, B. B. (1999). Ethical issues in risk communication: Continuing the discussion. *Risk Analysis, 19*(3), 348–355.

Johnstone, C. L. (1981). Ethics, wisdom, and the mission of contemporary rhetoric: The realization of human being. *Central States Speech Journal, 32*, 177–188.

Johanneson, R. L. (1996). *Ethics in human communication* (4th ed.). Prospect Heights, IL: Waveland Press.

Martinelli, K. A., & Briggs, W. (1998). Integrating public relations and legal responses during a crisis: The case of Odwalla, Inc. *Public Relations Review, 24*(3), 443–460.

Mileti, D. S., & Sorenson, S. (1990). *Communication of emergency public warnings: A social science perspective and the state-of-the-art assessment.* Oak Ridge, TN: Oak Ridge National Laboratory, Department of Energy.

Nilsen, T. R. (1974). *Ethics of speech communication* (2nd ed.). Indianapolis, IN: Bobbs-Merrill.

Phillips, R. (2004). Some key questions about stakeholder theory. *IVEY Management Services*. Retrieved December 10, 2007, from http://www.iveybusinessjournal.com/view_article.asp?intArticle_ID=471

Rowland, R., & Womack, D. (1985). Aristotle's view of ethical rhetoric. *Rhetoric Society Quarterly, 15*, 13–31.

Seeger, M. W. (1997). *Ethics and organizational communication.* Cresskill, NY: Hampton

Seeger, M. W., & Hipfel, S. (2007). Legal versus ethical arguments: Contexts for corporate social responsibility. In S. May, G. Cheney, & J. Roper (Eds.). *The debate over corporate social responsibility* (pp. 155–167). New York, NY: Oxford.

Sellnow, T. L., & Ulmer, R. R. (1995). Ambiguous argument as advocacy in organizational crisis communication. *Argumentation & Advocacy, 31*, 138–150.

Sellnow, T. L., Ulmer, R. R., & Snider, M. (1998). The compatibility of corrective action in organizational crisis communication. *Communication Quarterly, 46*, 60–74.

Singer, P. (2002). *Animal liberation.* New York: Harper Collins.

Singer, P., & Mason, J. (2006). *The way we eat: Why our food choices matter.* Kuntztown, PA: Rodale Institute.

Ulmer, R. R. (2001). Effective crisis management through established stakeholder relationships: Malden Mills as a case study. *Management Communication Quarterly, 14*, 590–615.

Ulmer, R. R., & Sellnow, T. L. (1997). Strategic ambiguity and the ethic of significant choice in the tobacco industry's crisis communication. *Communication Studies, 48*, 215–233.

Ulmer, R. R., & Sellnow, T. L. (2000). Consistent questions of ambiguity in organizational crisis communication: Jack-in-the-Box as a case study. *Journal of Business Ethics, 25*, 143–155.

Ulmer, R. R., Sellnow, T. L., & Seeger, M. W. (2007). *Effective crisis communication: Moving from crisis to opportunity.* Thousand Oaks, CA: Sage.

Zwart, H. (2000). A short history of food ethics. *Journal of Agricultural and Environmental Ethics, 12*, 113–126.

Chapter 12
Future Directions

> *Society would be safer, smarter, and fairer if our organizations and their masters could admit their limitations, declaring frankly that they cannot control the uncontrollable.*
>
> —Lee Clarke (1999, p. 137)

As we shift to a prospective outlook on risk communication, the future is fittingly uncertain. With the passing of each day, we advance in our understanding of existing risks. Conventional wisdom among many risk theorists is that modern life is much safer than ever before. Modern conveniences reduce risks in many ways. Medical techniques ensure longer lives for many people. Public health efforts have significantly reduced the threat of many diseases. Government regulation in developed nations protects the public against tainted food, faulty products, poor construction techniques, and dangerous medications. These advances, however, are compromised by continuous changes that generate risks heretofore never considered. In other words, the world is simultaneously a safer and more dangerous place today than it was yesterday.

In this chapter, we address the future directions of risk communication. We first introduce normal accidents theory (NAT) and chaos theory perspectives for revealing the unanticipated risks caused by the same safety measures we employ to reduce risk. Next, we discuss the potential for abusing and exploiting the message-centered approach to risk communication. We then offer the Center for Disease Control and Prevention's Crisis and Emergency Risk Communication (CERC) model as an appropriate application of the concepts covered in this book. We conclude the chapter with a discussion of interesting cases that challenge our current assumptions about risk communication.

Normal Accidents Theory (NAT)

Technological advancement is not synonymous with risk reduction. Perrow (1999) observes that high risk technology is multiplying, thereby creating a greater potential for catastrophes whose potential impact range is ever-increasing. Perrow identifies tight coupling and interaction as two key factors leading to the rising threat of technology. *Tight coupling* refers to the lack of slack between or among systems. When tight coupling is present, one system has the potential to impact another. For

T. L. Sellnow et al. *Effective Risk Communication*

DOI: 10.1007/978-0-387-79727-4_12

example, factories and power plants are now built closer to cities. Thus, those living in the community near a factory or power plant are at a greater risk of injury from malfunctions such as explosions or the emission of hazardous chemicals in the air. *Interaction* refers to the cumulative effect of multiple failures leading to an accident. For example, no single failure in leadership or communication deprived Hurricane Katrina victims of the food and water they needed to survive in the city. Rather, multiple failures by multiple agencies resulted in citizens of New Orleans suffering needlessly in the hurricane's wake. Because tight coupling and the ensuing interactions are prevalent in current technological advances, Perrow contends that events such as Hurricane Katrina are best described with the seemingly paradoxical term: normal accidents. Specifically, Perrow defines *normal accidents* as "the interaction of multiple failures that are not in a direct operational sequence" (p. 23).

Perrow (1999) explains that the interaction leading to normal accidents is non-linear; that is, no single cause is typically to blame for a crisis. For example, the Milwaukee *Cryptosporidium* contamination reflects multiple, seemingly unrelated failures. The water should not have been contaminated, but when it was the filtration system should not have failed. Once the water was contaminated and the filtration failed, public health officials should have advised the city's leadership to warn those with compromised immune deficiency that they were at greater risk. Thus, the *Cryptosporidium* case resulted from the non-linear confluence of errors by two agencies that, prior to the crisis, had only a passing association. Once the water filtration technology failed, public health officials in turn failed to issue a proper warning. Consequently, Milwaukee's leadership was guilty of ignoring a segment of its population during a crisis that should have been averted by the normal working system of the city's fresh water supply.

Reflecting on Perrow's work, Busby (2006) agrees that "the interactive complexity and tight coupling–and the contradictory requirements they impose on organizational decision making–make accidents virtually inevitable" (p. 1376). Such was the case in the Odwalla crisis. Odwalla was known as a producer of beverages designed for the health-conscious person. The fact that the company did not pasteurize its juice made its products an alternative with potentially higher quality and nutritional content. The lack of pasteurization created the type of contradictory requirement Busby identifies. The lack of pasteurization increased nutrition, but it also heightened the risk that the product could harbor dangerous pathogens. Odwalla's assumption that the acid in the product would make it safe proved to be an ill-fated compromise. Unfortunately, competition in the market place constantly exposes organizations to such contradictory requirements. For example, producing pet food at a lower price is appealing to customers. Doing so, however, may require additional risks for companies as they move to cheaper providers of bulk ingredients.

Perrow (1999) further warns the steady increase in high risk technology is not likely to level off soon. He argues that "shoddy construction and inadvertent errors, intimidation and actual deception" are "part and parcel of industrial life" (p. 37). The result of such failings may produce only minor problems, but the interaction of these minor problems can lead to major disasters. As Perrow contends, "It is in the minor abnormalities . . . that the system accident is spawned" (p. 57). From the perspective of NAT, many modern systems reach a level of size and complexity in their routines, procedures, technology, communication, and functions specialization where the system can no longer sufficiently recognize minor abnormalities in time to anticipate and avoid catastrophic failures. In the case of the 1979 accident at Three Mile Island, for example, a series of relatively small accidents and failures combined to create the worst nuclear accident in U.S. history. As food production has become more global and more technologically sophisticated, the probabilities of normal accidents have increased. Cases such as the melamine-tainted pet food of 2007 and the outbreak of hepatitis A at Chi Chi's restaurants can be understood as normal accidents.

NAT provides a framework for understanding the rising uncertainty and evolving risk associated with new technology. Some accidents, such as plane crashes and bridge collapses, are so complex that, even after years of post-crisis analysis, a single, clear cause is never identified. NAT also reveals a potential abuse of the message-centered approach to risk communication. Clarke (1999), using the foundations of NAT, admonishes stakeholders at all levels to beware of industry's exploitation of risk messages.

Chaos Theory

The series of potential relationships identified in NAT is incomprehensible. Nevertheless, potential patterns of problems can be observed through a mindful approach. Chaos theory offers an explanation for how such complex patterns of unanticipated interaction can be identified before, during, and after a crisis. Like NAT, chaos theory is based on the assumption that causal logic, linear proportionality, and reductionist methods are woefully inadequate for understanding the countless ways the discrete features of tightly coupled systems may interact (Kauffman, 1993; Keil, 1994; Matthews et al., 1999). As an alternative to such obvious causal thinking, chaos theory emphasizes such concepts as the impact of small variance and the frequency of unanticipated outcomes (Sellnow & Seeger, 2001). Chaos theory has been applied to risk and crisis communication as a means for understanding both the unanticipated disruptions of existing systems and the synergistic reconstruction of those systems (Murphy, 1996; Sellnow et al., 2002).

For purposes of understanding the function of risk communication, chaos theory can be dissected into four conceptual elements: bifurcation, fractals, strange attractors, and self-organization (Murphy, 1996; Seeger et al., 2003). We describe each of these elements in the following paragraphs.

Bifurcation

Simply put, bifurcation is the disruption of order. In a moment of calamity, assumed patterns of order are lost, leaving individuals momentarily confused, disoriented, and frustrated. Thus, bifurcation results in radical change after which "previous assumptions, methods, patterns, and relationships can no longer function" (Seeger et al., 2003, p. 31). Weick (1993) describes bifurcation as a cosmology episode where existing forms of sensemaking fail to account for the unforeseen experiences. The human reaction to such episodes is: "I've never been here before, I have no idea where I am, and I have no idea who can help me" (pp. 634–635). This loss of meaning is overcome through fractals.

Fractals

Fundamentally, fractals are units of measurement that allow us to identify patterns of organization. Thus, any effort to predict and control risk levels is essentially based on the observation of fractals. As we seek to observe patterns, the patterns we identify are profoundly influenced by the degree of sophistication in our observation of fractals. Murphy (1996) argues that "concentration on individual units can yield insignificant or misleading information" (p. 99). She notes that fractal observation should include multiple interactions. For example, Mandelbrot (1977) observed that viewing coastlines in 100 miles versus 10 miles increments provided an entirely different set of conclusions. In short, fractals provide the evidence upon which organizations base their decisions.

Strange Attractors

In their most basic essence, strange attractors are those values, principles, and social assumptions upon which order is based. The disruption produced by bifurcation demands that existing biases, tensions, and conflict be reconsidered in an effort to restore order. As Seeger et al. (2003) explain, "bifurcation creates a moment in which the status quo is suspended and established relationships are amenable to a fundamental reordering" (p. 34). They observe that, individuals and organizations that may not have cooperated in the past may unite because "the attraction of common threat throws groups together and demonstrates commonalities and overarching goals" (p. 34). In short, the disruption to order may foster alliances that are novel to the system. In some cases, adversaries may find themselves working together in the self-organization process.

Self-Organization

Ultimately, order is restored to a system in the form of self-organization. Self-organization is a naturally occurring process through which systems realign in the wake of bifurcation (Sellnow et al., 2002). As strange attractors unite, "new forms, structures, procedures, hierarchies, and understanding emerge, giving a new form to the system, often at a higher level of order and complexity" (Sellnow et al., p. 272). This higher level of complexity, in most cases, enables the organization to better manage the risk issues that preceded the bifurcation.

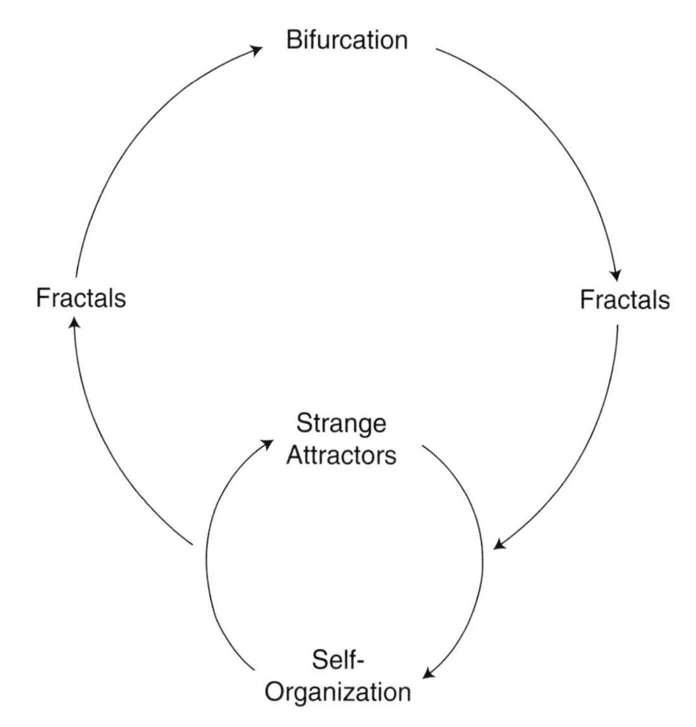

Fig. 12.1 Chaos model

Figure 12.1 depicts the interaction of the chaos elements described above. The model begins with bifurcation. As Perrow (1999) explains, such disasters are often created when organizations fail to mindfully consider all of the available evidence. Only after the disaster or bifurcation, is the problematic interaction of two elements observed. Thus, bifurcation leads, first, to fractals. The disaster reveals previously unknown or unattended evidence. Next, strange attractors are drawn together in a process of self-organization. Order is restored through this ongoing interplay between relevant entities, their value structures, and their fundamental objectives. In this interaction, risk communication is paramount. The relevant organizations and their leaders must determine the cause of the bifurcation and take steps to avoid such disasters in the future. Finally, the system returns to fractals. As such, the new order must continue to monitor its environment in hopes of discovering problematic interactions before they lead to another episode of bifurcation. If, however, risk communication and risk monitoring fail, the system returns to bifurcation and the process repeats itself.

Chaos theory has been applied to risk and crisis communication in a variety of settings. Freimuth (2003) used chaos theory to describe the communication challenge in the self-organization process at the CDC following the anthrax attacks in 2001. Sellnow et al. (2002) applied chaos theory to the risk communication strategies and subsequent recovery following a major flood. McIntyre (2007) observed the chaotic elements of the communication process in responding to a train derailment that released a plume of anhydrous ammonia over the city of Minot, North Dakota. Chaos theory was used in these studies to reveal unforeseen weaknesses in an existing system and the ultimate recovery of that system. In each of these applications the authors note that the organizations eventually established methods for monitoring and discussing relevant fractals more closely. The hope is that such observation will enable organizations to more readily recognize points of converging evidence that a problem is looming. By doing so, organizations are more likely to disrupt a potential interactions that could lead to bifurcation.

Potential Abuse of a Message-Centered Approach to Risk Communication

A message-centered approach to risk communication encourages individuals and organizations to generate, collect, and evaluate multiple risk messages from a range of perspectives and to base decisions on converging information. Regardless of how much convergence exists on a given risk issue, there is always some degree of uncertainty. When individuals and organizations fail to admit this uncertainty, even when convergence is strong, they have the potential to mislead others. Clarke (1999) makes an elaborate case against organizations that make misleading and absolute claims about managing a risk issue. He explains that when organizations publicize documents explaining their plan to manage risk they are engaging in a symbolic or message-centered process. Clarke points out that such documents "are rhetorical instruments that have political utility in reducing uncertainty for organizations and experts" (p. 13). When these documents speculate beyond the convergence of clear evidence and experience, they became what Clarke calls "fantasy documents" (p. 13).

Organizations commonly generate management plans that are designed to minimize risk. In fact, we list this process among our best practices for effective risk communication. The problem occurs when organizations create and oversell such plans even when "the knowledge and experience necessary to know what would make for a realistic plan are unavailable" (p. 14). Clarke labels such ill-conceived plans "fantasy documents" (p. 14). The harm of fantasy documents is in their capacity to mislead or manipulate stakeholders by claiming to provide a deceptive level of security. Perrow (1999) argues that such deception is based on an unwillingness to accept or admit to the "mystery" that lingers in many aspects of industries such as nuclear power (p. 29).

Clarke (1999) offers the Exxon *Valdez* accident as a convincing example of how industries use fantasy documents to persuade stakeholders to accept questionable risks. Before Exxon was allowed to transport oil through the pristine waters of Valdez, Alaska, the company was required to demonstrate its ability to immediately address and mitigate any spill that could occur in the area. Exxon composed an elaborate plan and warehoused spill management equipment worth thousands of dollars in the area. If the plan was any better than a fantasy at the outset, it certainly was woefully inadequate as time passed. The plan and the equipment quickly became dilapidated over time. When the spill occurred, Exxon lacked the equipment, personnel, and expertise to address the spill with immediacy. Exxon's cleanup plan was exposed as a fantasy. To date, debate over the cleanup continues and Exxon has spent over a billion dollars in the process.

The threat produced by fantasy documents is that consumers, workers, and investors will be duped into assuming a product or production process is safe when, in actuality, an industry or organization is engaging in high-risk behavior. Enron convinced investors that it was making money while, in reality, the company was experiencing millions of dollars in losses annually. Jack-in-the-Box restaurants argued that their grill temperatures exceeded the federal minimum when, in reality, the company's grills did not meet Washington state's standards. When the fast-food chain was responsible for a widespread *E. coli* outbreak due to undercooked hamburgers, its crisis management plan was revealed as utterly inept. These examples support Clarke's (1999) observation that "when fantasies are proffered as accurate representations of organizational capabilities, we have the recipe not only for organizational failure but for massive failure to the publics" (p. 168). The primary argument is clear: risk communication must not outpace the legitimate capacity for an organization, agency, or industry to respond fully to crises.

Thus far, we have discussed the increasing complexity of risk and the temptation of organizations and industries to overstate their claims regarding their preparation and capacity for managing risk. Clarke (1999) argues that the best means for addressing the gap between capacity and need is "the forthright admission that risk and danger are being created" along with "an honest assumption of the costs of creating danger, as well as an honest appraisal of the uncertainties our organizations create" (p. 171). Within this context of honesty, all stakeholders can engage in a dialogue that produces convergence on the best risk management options available. The CDC has dedicated considerable time and resources to developing a strategy for meeting this objective. This project is summarized in their CERC model. In the next section, we detail this model and describe how it can serve as a future direction in risk communication.

Crisis and Emergency Risk Communication Model

One of the responses to the challenges described above is the CERC framework developed by the CDC. This model (see Fig. 12.1) is an effort to develop a comprehensive and integrated framework for communication both about risks before they emerge and about crises after they are triggered (Reynolds & Seeger, 2005; Reynolds & Seeger 2007; Reynolds et al., in press). As an integrative framework, the CERC model is an effort to coordinate strategies of risk and crisis communication (Courtney et al., 2003). Those involved in its creation established a noble, if audacious, goal for the CERC model:

> [T]o provide information that allows an individual, stakeholders or an entire community, to make the best possible decisions about their well-being, under nearly impossible time constraints, and to communicate those decisions, while accepting the imperfect nature of their choices. (CDC, 2007, para 1)

Notice that the CERC model makes no promises about being able to control any risk or to avoid any crisis. Rather, the model is designed to approach risk and crisis as an inevitable part of life and to empower stakeholders. Rather than fantasy, the model is rooted in the reality that the potential for emerging risks to fester into disastrous crises is rising, not receding.

As political tensions and global trade evolve, there are ever increasing "challenges for the medical and public health community to communicate in accurate, credible, timely, and reassuring ways" (Reynolds & Seeger, 2005, p. 45). Reynolds & Seeger (2005) argue that to address the future challenges in risk communication, we "must be strategic, broad based, responsive, and highly contingent" (p. 49). They contend that the CERC model meets this need. In October 2002, the CDC initiated an education program designed for public health officials (Reynolds et al., 2002). CERC was developed "primarily as a tool to educate and equip public health professionals for the expanding communication responsibilities of public health in emergency situations" (Veil et al., in press). The model is based on an extensive review of existing risk communication and crisis planning literature, as well as ongoing research conducted by and for the CDC. This inclusive nature of the CERC model allows us to generalize beyond public health to include all risk planning where public safety is involved.

CERC is a response to a general recognition that an integrated and coordinated risk/crisis approach is more effective than either approach alone. A number of high profile events, including the 2001anthrax episode and Hurricane Katrina, illustrated that risk messages before an event have an important impact on how the public responds to an event. There are five general stages of CERC: pre-crisis, initial event, maintenance, resolution, and evaluation. In each stage, a set of recommended communication strategies is described (see Fig. 12.2). During pre-crisis, for example, communicators should use traditional risk communication strategies. These activities include educating the public about risks and appropriate responses to avoid risks including changing behaviors so that risks are reduced. In the crisis stage, communication activities such as information about the nature of the threat, responses by officials, and recommendations for self-efficacy or individual response may be required. As the crisis moves into the resolution stage, communicators face a different set of exigencies. At this point, more detailed information about the nature of the risk is needed. During the resolution stage, questions of cause, blame, and responsibility for the risk typically emerge. These questions, then, require official or organizational

I. Pre-Crisis (Risk Messages; Warnings; Preparations)
Communication and education campaigns targeted to both the public and the response community to facilitate:
- Monitoring and recognition of emerging risks
- General public understanding of risk
- Public preparation for the possibility of an adverse event
- Changes in behavior to reduce the likelihood of harm (self-efficacy)
- Specific warnings messages regarding some eminent threat, such as evacuation notices, take shelter warnings, product recalls, etc.
- Alliances and cooperation with agencies, organizations, and groups
- Development of consensual recommendations by experts and first responders
- Message development and testing for subsequent stages

II. Initial Event (Uncertainty Reduction; Self-Efficacy; Reassurance)
Rapid communication to the general public and to affected groups seeking to establish:
- Empathy, reassurance, and reduction in the public's emotional turmoil
- Designated crisis/agency spokespersons and formal channels and methods of communication
- General and broad-based understanding of the crisis circumstances, nature of the threat, consequences, and anticipated outcomes based on available information
- Reduction of crisis related uncertainty
- Specific understanding of emergency management and medical community responses
- Understanding of self-efficacy and personal response activities (how/where to get more information; check on neighbors; avoid contaminated water, etc)

III. Maintenance (Ongoing Uncertainty Reduction; Self-Efficacy; Reassurance)
Communication to the general public and to affected groups seeking to facilitate:
- More accurate public understandings of ongoing risks
- Understanding of background factors and issues
- Broad based support and cooperation with response and recovery efforts
- Feedback from affected publics and correction of any misunderstandings/rumors
- Ongoing explanation and reiteration of self-efficacy and personal response activities (how/where to get more information) begun in Stage II.
- Informed decision-making by the public based on understanding of risks/benefits

IV. Resolution (Updates Regarding Resolution; Discussions about Cause and New Risks/New Understandings of Risk)
Public communication and campaigns directed toward the general public and affected groups seeking to:
- Inform and persuade about ongoing clean-up, remediation, recovery, and rebuilding efforts
- Facilitate broad-based, honest, and open discussion and resolution of issues regarding cause, blame, responsibility, and adequacy of response
- Improve/create public understanding of new risks and new understandings of risk as well as new risk avoidance behaviors and response procedures
- Promote the activities and capabilities of agencies and organizations to reinforce positive corporate identity and image

V. Evaluation (Discussions of Adequacy of Response; Consensus about Lessons and New Understandings of Risks)
Communication directed toward agencies and the response community to:
- Evaluate and assess responses, including communication effectiveness
- Document, formalize, and communicate lessons learned
- Determine specific actions to improve crisis communication and crisis response capability
- Create linkages to pre-crisis activities (Stage I)

Reynolds, B. & Seeger, M. W. (2005). Crisis and Emergency Risk Communication as an integrative model. *Journal of Health Communication Research, 10*(1), 43–55.

Fig. 12.2 Crisis and emergency risk communication model

responses with a broad-based, honest, and open discussion of the risks. Finally, the evaluation stage is an opportunity to create and communicate the lessons from the crisis. Evaluation creates the opportunity for health communicators to prepare for the next infectious disease or food borne illness outbreak.

CERC is a tool for risk communicators to use in an effort to more effectively communicate about those risks that may eventually emerge into a crisis. By integrating and coordinating the strategies of risk communication and crisis communication, the probabilities of facilitating effective understandings of these issues on the part of both decision makers and the public is increased and the probability of falling victim to a fantasy document is diminished.

The Challenge of Multiple Audiences

The CERC model offers a refreshing alternative approach to address the increasing complexity of risk situations. As we have argued throughout this book, however, convergence and planning in risk situations must account for the multiple audiences. The complexity of crisis situations is undeniable. Our decision to use the case study approach to reveal the multidimensional aspects of this complexity serves as evidence for why multiple points of view are needed when seeking enhanced knowledge of contemporary events involving crisis and risk. As a result of this complexity, scholars and practitioners seeking to understand risk are placed in a difficult situation because no matter how they try to find convergence in the process of persuading audiences to respond in a particular way, they cannot control how their risk messages will be perceived. However, risk communicators can control how their messages are created and presented, and that should be part of the focus for future research in risk communication. Three audience-centered perspectives have been introduced in this book to explain approaches taken by risk communicators in their effort to create and present messages. These include the *culture neutral* approach, the *culturally sensitive* approach, and the *culture-centered* approach.

As presented, the culture neutral approach is the default position for most risk communicators who seek to reach the general public. These risk communicators believe that if they present the facts as they exist, the general public, being rational and responsible, will accept the information as presented and act upon it. This elitist view is reflected in the single spokesperson model where the message is scripted and presented to all audiences simultaneously. These risk communicators consider themselves to be culture neutral because they consciously or unconsciously do not acknowledge cultural differences in their messages; everyone gets the same information in the same manner. However, the risk communicator who does not acknowledge diverse publics or gaps in the interacting arguments may be perceived by particular audiences as being *culturally insensitive.* Such was the case with ConAgra's initial response to its *Salmonella* warning. The Hurricane Katrina case study is a powerful example of how messages presented to the general public were culture neutral at best and culturally insensitive in fact. In the Hurricane Katrina case, the absence of messages creating convergence for the African American community and those of lower socio-economic status illustrated the need for a different approach from the culture neutral strategy used by emergency managers in that situation. Had the risk messages been constructed and presented differently, the crisis might have had a different outcome for thousands of people who faced the aftermath of Hurricane Katrina.

For some risk communicators, the culturally sensitive approach reflects progress in acknowledging the presence of different publics within the general audience receiving a risk message. As risk communicators seek to accomplish the objective of persuading people to respond in an appropriate way to their message, they identify certain cultural characteristics of their target audiences as markers that will make them to appear more culturally sensitive. The communicators may know that the audience is more likely to pay attention if there are visual signifiers present (flags, symbols) or auditory markers (sirens, warning sounds). As a result, they include these signifiers and markers in their messages. They may know that particular groups prefer to receive messages from members of their own cultural group. Consequently, they build relationships with cultural agents who will present the scripted message in times of crisis. They may know that literacy levels vary among groups, resulting in messages that are presented in simpler, less complicated terms. All of these strategies help to create and present messages to the target audience as the audience prefers them. However, despite the culturally sensitive approach, the message is still the same and the risk communicator is still in control of what the target audience should be concerned about and what the preferred responses and behaviors should be.

While being culturally sensitive is preferred over the culture neutral or culturally insensitive approach, problems remain. In the tainted Odwalla apple juice case, the company utilized a culturally sensitive approach by stepping up its communication with the media and offering consumers multiple ways to remain safe. While the messages were designed for a more affluent group of consumers, the company was apologetic, compensatory, and willing to undergo a safety assessment of its product and the process used to prepare it for public consumption. While focusing on its market segment with strategies that appeared compassionate, honorable, and truthful, it still reflected the scripted message of the company: that it could be trusted, the juice industry was worthy of continued support, and the public should remain faithful customers. There appeared to be no effort to involve the public in any dialogue about the nature of the crisis or how the risk might be communicated across socio-economic groups.

In contrast, the culture-centered approach allows for the greatest impact by involving cultural groups in the process of discussing potential crises and the risks associated with them. The culture-centered approach removes the risk communicators from the elite position of determining what is good or best for the different cultural groups. The varied perceptions of crisis are accounted for and risks associated with these crises are discussed in real terms that have meaning for the groups involved. This culture-centered approach will alter the way risk communicators develop their messages because the cultural groups will determine what form of reasoning is most persuasive. In addition, the way risk messages are presented using a culture-centered approach will change. Instead of a single spokesperson or organization controlling the release of information, the groups constituting the public will share in the responsibility for determining the nature and timing of the information's dissemination. The Milwaukee *Cryptosporidium* crisis case serves as an example of how a culture-centered approach could have made a difference in the way the situation was resolved. The regulatory powers in Milwaukee did not acknowledge community concerns about the water quality. By failing to engage in open, honest, and timely communication, they excluded the public from the process. By the time the officials acknowledged the crisis, it was too late. Including all groups in the dialogue about the importance of water quality and the role citizen groups play in helping to identify and disseminate information about the crisis might have resolved the situation earlier.

The culture neutral (insensitive), culturally sensitive, and culture-centered approaches to risk communication are strategies used by risk communicators when preparing and disseminating information in pre-crisis, crisis, and post-crisis situations. Additional study is needed as we move forward in making risk communication more responsive to the needs of those who receive these messages. For example, are there times when information should be presented in a culturally neutral way? If the single spokesperson model is used, is it possible for culturally sensitive or culture-centered communication to occur? In times of crisis, is there time to be culturally sensitive or should spokespeople be more concerned about the utility of getting the message out to the general audience? While we want to be sensitive to the perspectives of the audiences who receive the messages, we must ask, "Is risk communication inherently dominated by an informed elite with the most information about the potential hazards and threats to human and animal life?" Finally, is the consideration of a culture-centered perspective feasible? The critical perspective that guides the culture-centered model suggests that information and power has been in the hands of the elites. The culture-centered model creates an opportunity for the voices of those who are not elites to be included in the dialogue about risk and crisis. In a pre-crisis context, such inclusion may be feasible. However, in a crisis situation, is time available for the voices of the underrepresented groups to be heard? Future study is needed as risk communicators seek to reconcile these and other perspectives with the best practices of risk communication.

Clearly, our risk communication must be flexible enough to address the uncertain nature of evolving risks and the potential impact of those risks on all stakeholders. In the next section, we offer examples of unusual or complex risk situations that have occurred recently. By doing so, we begin to characterize some of the future directions risk communicators must consider.

Challenging Cases

As Perrow (1999) explains, new forms of unanticipated risk are constantly emerging. The following cases illustrate this evolution. For each case, we identify a liability that produces challenges for effective risk communication.

The Liability of Cultural Conflict

As Japanese citizens prepared to celebrate the new year, they were greeted with a perplexing food-borne outbreak linked to *gyozo*, translated as dumplings, produced in China. The Japanese government confirmed 10 cases of illness, and as the story developed in the Japanese media, 4,000 concerned individuals contacted health authorities with inquiries or reports of illness. Making matters worse, one of the bags of dumplings was reported to have been corrupted with small holes approximately the size needed for a syringe needle. Questions of both intentional contamination and incompetence by the Chinese government raged in the Japanese media. Methamidophos, a substance banned in both China and Japan was found in the dumplings. Japanese officials threatened to boycott Chinese products, while China insisted that an investigation of its factories revealed no sign of methamidophos. As the two countries engaged in a war of words, Japanese consumers were left with limited information and little assurance that their risk was being addressed. Lingering animosity between China and Japan hindered the risk communication process. Thus, this case emphasizes the need for risk communicators to further assess the complications of culture in managing risk issues in a global market.

The Liability of Shifting Blame

As we discussed in Chapter 9, ConAgra issued a statement that shifted a portion of the responsibility for the outbreak to consumers. In their first press release, ConAgra stated that it believed the issue was related to the *consumer* not cooking the pies properly. Working with the USDA, ConAgra Foods chose to revise the cooking instructions located on the food packages. ConAgra's decision to shift blame to the consumers represents an ethical challenge. Naturally, consumers should prepare their food safely. Still, the fact that ConAgra sold a product contaminated with *Salmonella* is problematic. Previous research indicates that shifting blame without absolute certainty that the organization does not bear some responsibility is ill-advised or unethical (Benoit, 1995; Hearit, 2006). For those who study and engage in risk communication, the question is, "At what point does the company's responsibility end, and the consumers' responsibility begin?" This is a difficult question that is made increasingly complex by the continuous evolution of technology and global marketing.

The Liability of Smallness

Despite a history of providing quality meat since 1940, Topps Meat Company closed its doors in the fall of 2007, six days after issuing the second largest beef recall in United States History. Thirty people in eight states were infected with

E. coli 0157:H7 bacteria strain after eating Topps hamburger patties. The company was eventually forced to recall nearly 22 million pounds of ground beef; although most had already been consumed.

Whittier Dairy Farms provided milk products for residents in the Boston area for a century before closing its doors permanently in January 2008. In the months before closing, listeria was identified in milk processed at the dairy. Sadly, the listeria contamination killed three men and one still-born baby. The financial toll of the recall and impending legal cases combined with market pressure was too much for the company to withstand.

These cases reveal two intensifying challenges for similar companies. Both Topps and Whittier Farms Dairy were relatively small by their industry standards. Despite stellar records of past service, both companies lacked the financial resources needed to endure the fallout from their crises. Without enormous financial reserves, organizations often cannot withstand the financial fallout from a major failure. Thus, there is little margin for error. Consequently, smaller organizations are extremely vulnerable to sabotage on a financial level. In other words, a single event caused by wrongdoers could be the undoing of a small or mid-sized company. As political tensions and the prevalence of terrorism continue to intensify, the potential for such attacks becomes a risk that warrants consideration.

The Liability of Mass Distribution

In the spring of 2007, a wave of illness in cats and dogs was linked to tainted pet food. Although dozens of brands were implicated in the illnesses, the problem was traced to contaminants in vegetable proteins imported from China. Melamine, a substance that is toxic to cats and dogs, was found in the vegetable protein. Including melamine in the product clearly represents an ethical violation. The greater and more challenging issue, from the perspective of risk communication, relates to the world's increasing reliance on mass distribution of products. The ethical lapse of a few individuals seeking to cut costs in one country can have a devastating impact on industries and consumers throughout the world. Thus, potential risk is magnified exponentially through the growing tendency of organizations to rely on a centralized distribution process. As Perrow (1999) argues, tight coupling, in this case centralized distribution, has the potential to magnify failures beyond our imagination. Thus, mass distribution results in an evolving and intensifying risk that poses a challenge for risk communicators.

The challenges described in this section reflect only a portion of the evolving issues that risk communicators must address in the future. Most importantly, these cases illustrate the increasing intricacy of risk communication and the importance of dialogue about planning in order to address this complexity.

Summary

As NAT and the emerging liabilities described above indicate, technology alone cannot adequately address the increasing complexity of risk in all levels of society. Effective risk communication is essential for mindfully collecting essential information and for generating the level of dialogue needed for creating an ethical strategy that adequately addresses the needs of all relevant parties.

As we advocate a mindful, representative, and ethical dialogue, however, we admit that achieving consensus in risk decision-making is often difficult, if even attainable. The uncertainty inherent in all risk discussion results in multiple positions and opinions that foster competing arguments. Our hope is that, through the strategies we advocate in this book, all parties can apply the best practices of risk communication to make decisions and form plans that are more mindful, inclusive, and ethical. Specifically, we encourage members of industry and stakeholders at all levels to observe the interaction of arguments on any given risk issue and identify points of convergence. Such convergence is the foundation for improved risk decision-making. The well-being of society depends on such prudent communication and planning.

References

Benoit, W. L. (1995). *Accounts, excuses, and apologies: A theory of image restoration strategies.* Albany, NY: State University of New York Press.

Busby, J. S. (2006). Failure to mobilize in reliability-seeking organizations: Two cases from the UK Railway. *Journal of Social Issues, 56*(1), 105–127.

Centers for Disease Control and Prevention. (2007). Retrieved February 22, 2008, from http://www.bt.cdc.gov/erc/cerc.asp

Clarke, L. (1999). *Mission improbable: Using fantasy documents to tame disaster.* Chicago: The University of Chicago Press.

Courtney, J., Cole, G., & Reybolds, B. (2003). How the CDC is meeting the training demands of emergency risk communication. *Journal of Health Communication, 8*, 128–129.

Freimuth, V. S. (2006). Order out of chaos: The self-organization of communication following the anthrax attacks. *Health Communication, 20*(2), 141–148.

Hearit, K. M. (2006). *Crisis management by apology: Corporate response to allegations of wrongdoing.* Mahwah, NJ: Lawrence Erlbaum Associates.

Kauffman, S. A. (1993). *Origins of order: Self organization and the nature of history.* New York: Oxford University Press.

Keil, L. D. (1994). *Managing chaos and complexity in government.* San Francisco: Josey-Bass.

Mandelbrot, B. B. (1977). *Fractals: Form, chance, and dimensions.* San Francisco: W. H. Freeman.

Matthews, M. K., White, M. C., & Long, R. G. (1999). Why study the complexity sciences in the social sciences? *Human Relations, 25*, 439–461.

McIntyre, J. J. (2007). Crisis narratives: Creating community order from chaos. Unpublished doctoral dissertation, North Dakota State University, Fargo.

Murphy, P. (1996). Chaos theory as a model for managing issues and crises. *Public Relations Review, 22*, 95–113.

Perrow, C. (1999). *Normal accidents: Living with high-risk technologies.* Princeton, NJ: Princeton University Press.

Reynolds, B., Hunter-Galdo, J., & Sokler, L. (2002). *Crisis and emergency risk communication.* Atlanta, GA: Centers for Disease Control and Prevention.

Reynolds, B., & Seeger, M. W. (2005). Crisis and emergency risk communication as an integrative model. *Journal of Health Communication, 10*, 43–55.

Seeger, M. W., & Reynolds, B. (2007). Crisis communication and the public health: Integrated approaches and new imperatives. In M. W. Seeger, T. L. Sellnow, & R. R. Ulmer (Eds.), *Crisis communication and the public health* (pp. 3–20). Cresskill, NJ: Hampton Press.

Seeger, M. W., Reynolds, B., & Sellnow, T. L. (in press) Crisis and emergency risk communication in health contexts: Applying the CDC model to pandemic influenza. In R. L. Heath (Ed.), *Handbook of risk and crisis and crisis communication.* New York: Routledge.

Seeger, M. W., Sellnow, T. L., & Ulmer, R. R. (2003). *Communication and organizational crisis.* Westport, CT: Praeger.

Sellnow, T. L., & Seeger, M. W. (2001). Exploring the boundaries of crisis communication: The case of the 1997 Red River Valley flood. *Communication Studies, 52*, 153–168.

Sellnow, T. L., Seeger, M. W., & Ulmer, R. R. (2002). Chaos theory, informational needs, and natural disasters. *Journal of Applied Communication Research, 30*, 269–292.

Veil, S. R., Reynolds, B., Sellnow, T. L., & Seeger, M. W. (in press). Crisis and emergency risk communication in health contexts: Applying the CDC model to pandemic influenza. *Health Promotion Practice.*

Weick, K. E. (1993). The collapse of sensemaking in organizations: The Mann Gulch disaster. *Administrative Science Quarterly, 38*, 628–652.

Index

Printed in the United States of America